Lenka Krpálková

Optimal growth of heifers and dairy herd production and profitability

AF138223

Lenka Krpálková

Optimal growth of heifers and dairy herd production and profitability

LAP LAMBERT Academic Publishing

Impressum / Imprint

Bibliografische Information der Deutschen Nationalbibliothek: Die Deutsche Nationalbibliothek verzeichnet diese Publikation in der Deutschen Nationalbibliografie; detaillierte bibliografische Daten sind im Internet über http://dnb.d-nb.de abrufbar.
Alle in diesem Buch genannten Marken und Produktnamen unterliegen warenzeichen-, marken- oder patentrechtlichem Schutz bzw. sind Warenzeichen oder eingetragene Warenzeichen der jeweiligen Inhaber. Die Wiedergabe von Marken, Produktnamen, Gebrauchsnamen, Handelsnamen, Warenbezeichnungen u.s.w. in diesem Werk berechtigt auch ohne besondere Kennzeichnung nicht zu der Annahme, dass solche Namen im Sinne der Warenzeichen- und Markenschutzgesetzgebung als frei zu betrachten wären und daher von jedermann benutzt werden dürften.

Bibliographic information published by the Deutsche Nationalbibliothek: The Deutsche Nationalbibliothek lists this publication in the Deutsche Nationalbibliografie; detailed bibliographic data are available in the Internet at http://dnb.d-nb.de.
Any brand names and product names mentioned in this book are subject to trademark, brand or patent protection and are trademarks or registered trademarks of their respective holders. The use of brand names, product names, common names, trade names, product descriptions etc. even without a particular marking in this work is in no way to be construed to mean that such names may be regarded as unrestricted in respect of trademark and brand protection legislation and could thus be used by anyone.

Coverbild / Cover image: www.ingimage.com

Verlag / Publisher:
LAP LAMBERT Academic Publishing
ist ein Imprint der / is a trademark of
OmniScriptum GmbH & Co. KG
Heinrich-Böcking-Str. 6-8, 66121 Saarbrücken, Deutschland / Germany
Email: info@lap-publishing.com

Herstellung: siehe letzte Seite /
Printed at: see last page
ISBN: 978-3-659-58200-4

Zugl. / Approved by: Prague, University of Life Sciences In Prague, Diss., 2014

Copyright © 2014 OmniScriptum GmbH & Co. KG
Alle Rechte vorbehalten. / All rights reserved. Saarbrücken 2014

Optimal growth of heifers and effect of milk yield level on dairy herd production, reproduction, and profitability

Lenka Krpálková[1]

Victor E. Cabrera[2]

Jindřich Kvapilík[1]

Jiří Burdych[1, 5]

Miloslava Štípková[1]

Peter Crump[3]

Luděk Stádník[6]

Mojmír Vacek[6]

[1]Department of Cattle Breeding, Institute of Animal Science, Přátelství 815, 10400 Prague 10 – Uhříněves, Czech Republic
[2]Department of Dairy Science, University of Wisconsin, 1675 Observatory Dr., Madison, WI 53716, USA
[3]College of Agricultural & Life Sciences, University of Wisconsin, 1675 Observatory Dr., Madison, WI 53716, US
[4] Department of Nutrition and feeding of farm animals, Institute of Animal Science, Přátelství 815, 10400 Prague 10 – Uhříněves, Czech Republic
[5]VVS Verměřovice s.r.o, Verměřovice 159, 561 52 Verměřovice, Czech Republic
[6]Department of Animal Husbandry, Faculty of Agrobiology, Food and Natural Resources, Czech University of Life Sciences Prague, Kamýcká 129, 165 21, Prague 6 – Suchdol, Czech Republic

Prague 2014

The authors thank the dairy producers who participated in this study. Their help was greatly appreciated. This research was supported by projects MZERO 0714 from the Ministry of Agriculture of the Czech Republic.

1 Introduction

Heifer rearing is an integral part of dairy herd turnover. The method of heifer rearing and length rearing period determine the cost and quality of heifers entering the dairy herd and replacing culled or problem cows. A well-raised heifer with a higher genetic potential is the best substitute for an inadequate cow and does not bring about any economic losses that might be incurred due to high rearing cost or poor performance of fresh heifers. The shortening of rearing period (non-productive time) would lead to cost reduction. This can be achieved by faster heifer growth, earlier onset of sexual maturity, breeding and calving.

Accelerated heifer growth, however, can be associated with excessive fat deposition in the udder or around reproductive organs and increased body condition, which may adversely affect performance and fertility of cows, and thereby their longevity. Heifer body maturity at first calving also influences milk yield, particularly in the first two lactations. The target first calving age should be set taking into account all the above-mentioned factors. In reality, however, the variation among herds in target first calving age and body weight of first bred heifers is great, and is associated with differences in weight gain and body condition levels among herds as well as different breed or individual specific traits in animals within one herd.

Dairy herds need enough replacement heifers. Therefore, the proper rearing management has a considerable impact on farm profitability. Nevertheless, the heifer rearing is often neglected in dairy herd management.

The aim of this thesis is to define consequences of heifer rearing management decisions, on the basis economic models of herd dynamics in biological context, developed for this purpose.

2 Literary review

2.1 Heifer growth and development

Growth is one of the basic processes that take place during the animal´s life. It is influenced by genetic potential, nutrition and environmental factors. During the development phase different body parts show uneven growth in relation to the whole body, so called growth allometry. After the nervous tissue development is completed, mainly the skeleton, muscle tissue and finally adipose tissue undergo further growth (Philips, 2010). According to Bar-Peled et al. (1997) until 3 months of age the body cells and parenchyma cells of the mammary gland develop at similar rates in the period from birth to 3 mo (months) of age, but the parenchyma of the mammary gland develops 3 to 4 times faster than the cells of the body in the period from 3 to 9 mo of age. Owens et al. (1993) suggested that the growth is completed when body maturity is achieved, the height and width diameters of the animal do not change any more and the volume of muscle tissue becomes stable.

If the growth is reduced due to nutrient deficiencies, afterwards there is a growth compensation, i.e. more intensive growth during the development phase, which can be beneficial as reported by Choi et al. (1997) and Park at al. (1998). Heifers that underwent the growth compensation had a lower proportion of adipose tissue in the mammary gland than those with no growth compensation. Ford and Park (2001) reported that if the growth compensation occurs during the allometric growth phase, the development of mammary gland is improved and energy and protein metabolism of the heifer are adjusted accordingly.

Brody´s , Gompertz´s or Richard´s functions can be used to describe and predict the heifer growth. Within these functions the age correction can be performed by linear interpolation. The functions are used to control the growth and manage automatic feeding systems correctly (Perotto et al. 1992; Bach a Kertz, 2010; Cue et al., 2012).

2.1.1 Feeding intensity and growth of heifers

Every farmer´s aim is to achieve maximum animal performance. Healthy and well- fed calves are the first prerequisite for excellent performance later in their lives (Zeman et al., 2006). Wattiaux (2011) reported that sexual maturity of heifers depends on their body weight rather than age. The sexual maturing starts when heifers reach 38 – 44 % of their adult body weight, i.e. usually at 270 – 300 kg. That is why the onset of sexual maturing is significantly influenced by the level of nutrition. Heinrichs and Gabler (2003) suggested that the age at first calving is influenced not only by circumstances around calving, but also by nutrition, health and environmental factors in the first 4 months of life.

Generally, farmers assume that if the first calving age is decreased, heifer raising costs are lower and their profit higher because of longer production period (Van Amburht et al., 1998; Mourits et al., 1999a). In order to decrease the age at first service, faster and more efficient heifer rearing is required. However, due to intensive feeding an excessive fat deposition in mammary gland tissue might occur, leading to reduced future milk production, mainly in the first lactation, and difficult conceptions (Drevjany et al., 2004; Daniels, 2010). Fat deposition in well-nourished heifers has an antagonist effect on milk production, causes depression and impairment of sexual functions or it can even lead to complete infertility. It can also result in an increased incidence of dystocia, ketosis, abomasum displacement, and consequently an increased culling rate (Hoffman et al., 1996; Le Cozler et al., 2008). According to Hansen et al. (2004) and Bicalho et al. (2007) heifers are facing a greater risk of being over-conditioned than cows, so are more likely to suffer of dystocia and other health problems. Heifers may become over-conditioned when the dietary ratio of energy and protein is inappropriate and/or when heat detection is not done properly, and consequently conception is postponed. Abeni et al. (2000) came to the same conclusion and they added that over-conditioned heifers have different plasma concentrations of the metabolites that are associated with energy, protein and mineral metabolism and liver function. Le Cozler et al. (2009b) concluded that the undesirable over-conditioning of pre-puberty heifers can have a long-lasting negative impact on cow performance.

On the contrary, with inadequate nutrition and low growth rates not only heifer raising costs are increased due to the rearing period extension, but also late sexual maturity onset, lower body weight at first calving and associated compromised calving ease (dystocia, stillbirths), and reduced first lactation milk yield (Le Cozler et al., 2008) can be expected. Dawson and Carson (2004) found that diets given to young heifers have a minor effect on future performance, although their results show that grazing has a favourable effect on the development of mammary gland and reduces lameness. Hoffman (2009) stated that low level of feeding if no corrections for some nutrients are made, usually does not lead to permanent fertility disorders in heifers, but it prolongs the pre-reproduction period considerably. In fact,

3

heifers have a great ability to compensate for periods of retarded grow by growth acceleration (Choi et al., 1997; Park at al., 1998; Dawson and Carson, 2004). In the study performed by Roberts et al. (2009) one group of heifers was fed *ad libitum* and the other one was subjected to feed restriction according to their body weight. The *ad libitum* fed heifers had better fertility results, but the economic analysis found a cost reduction by $33 per pregnant heifer in the restricted group. This implicates that restricted feeding in the post-weaning period can be cost-effective.

Heifers do not have constant growth rates. In fact, the most usual pattern is a faster growth between the birth and onset of puberty (6-10 months of age), followed by a slower growth period. Therefore, the most efficient way of increasing heifer growth potential is to improve nutrition in the per-puberty phase (Heinrichs and Gabler., 2003; Bouška et al., 2006). Mourits et al. (1999a), Abeni et al. (2000) and Shamay et al. (2005) suggested that the heifer rearing period (particularly in terms of nutrition and feeding management) should be divided into two consecutive phases, i.e. before and after the onset of sexual maturity. Le Cozler et al. (2008) reported that even prenatal feeding management has an impact on future performance and profitability of heifers.

2.1.2 Nutrition and growth of heifers during the pre-puberty period

Even in an early calf's life, the rearing outcome can be improved by providing the optimum amount of high quality colostrum to the calf. It was found that in the first 2 months of life the heifer's growth is positively correlated with her future milk performance (Bach and Kertz, 2010). Most researchers agree on the fact that milk is better for calves than milk replacer, but they admit that a higher cost of waste milk feeding can be a limiting factor for its use. Milk-fed calves grow faster, easily transfer to vegetable feed after the weaning and achieve puberty earlier. Moreover, a lower incidence of scours was observed (Foldager and Krohn, 1994; Jasper and Weary, 2002; Le Cozler et al., 2008; Uys et al., 2011). Shamay et al. (2005) drew a similar conclusion and also reported that milk fed calves had higher weight gain, which however did not influence the skeletal growth, and had higher milk fat in the first lactation. According to Doležal et al. (2001) it is necessary to provide greater amounts of milk replacer daily to achieve a higher growth rate, and wean calves gradually. Blome et al. (2003) and Drevjany et al. (2004) reported that calves weaned too early showed a retardation of growth and did not take advantage of quite an expensive milk replacer. A high protein content in milk replacer should provide for faster growth of muscle and reduced fat deposition. Veauthier et al. (2000) and Spiekers et al. (2009) suggested that also milk replacer protein quality is important because microbial protein only constitutes a small part of the calf´s protein nutrition.

Bouška et al. (2006) suggested that milk replacer should have a high level of crude protein (up to 28 %), ensuring weight gain of 650 – 700 g/day. In order to provide full value nutrition to the calf, a special calf starter should be provided as early as a few hours after the birth to stimulate the rumen wall and papillae development and formation of volatile fatty acids that initiate microbial activity supporting forestomach development (Nordlund et al., 1999; Doležal et al., 2001). The starter feed should be presented twice a day, at least 1.5 kg at each meal, along with appropriate amounts of hey (1.5 - 2 kg), (Bouška et al., 2006). Crude protein content in the starter feed should range between 18 – 20 % (Doležal et al., 2001). Kertz et al. (1984) and Heinrichs and Gabler (2000) stated that the calves which do not receive water up to their requirement, have a lower starter feed intake and retarded rumen development.

From about 2 to 3 months of age, calves can be offered free choice total mixed ration (TMR) consisting of silage and concentrate, with crude protein level of about 16%. The dietary nutrients should be adequate to average daily weight gain (ADG) of 850 – 900 g (Bouška et al., 2006). The aim of the first heifer rearing phase is to achieve fast growth, large body frame and good development of the udder. Increased heifer growth rate requires higher crude protein levels in the diet (Daccarett et al., 1993). According to Heinrichs and Gabler (2003) it is important that pre-puberty Holstein heifers would be able to use digestible protein and energy efficiently. Thus, heifer raising costs are reduced and life-time herd production is improved. Bouška et al. (2006) stated that if the body size of heifers is inadequate, the diet is likely to contain too low percentage of crude protein or non-degradable nitrogen. According to Bethard et al. (1997) heifer requirements for energy and nutrients change during the rearing period.

A young heifer has a limited feed intake capacity but high requirements for dietary proteins and minerals. According to Lammers and Heinrichs (2000) the requirements for digestible protein in growing heifers from 2 to 6 months of age and over 6 months are 16% and 12%, respectively. Energy requirements for heifer growth are decreasing linearly from 2 months of age on. On the basis of their study results they calculated the ratio of digestible protein to metabolisable energy as 49-51 : 1 g/Mcal for heifers from 6 to 12 months. Heinrichs and Gabler (2003) found that the protein to energy ratio is suitable for the evaluation of protein requirements. The lowest protein to energy ratios of 49.3 and 52.2 : 1 resulted in fast growth and high feed efficiency. For the ratio of 49.3 : 1 the highest nitrogen efficiency was observed. Gasser et al. (2006) suggested that an increased energy intake in early weaned heifers from 126 to 196 days of age would lead to an earlier onset of puberty, regardless of diet. However, according to Le Cozler et al. (2009a) an excessive growth from 120 to 300 days of age (approximately 150 - 260 kg of body weight) is unfavourable for the future milk production. Abeni et al. (2000) pointed out that ADG of 700 g - 800 g is ideal to achieve target milk performance, but between 90 and 300 kg of body weight ADG over 700 g is not favourable for the mammary gland growth.

In practice, this relationship indicates how important it is to control the growth until sexual maturity is reached. From the onset of sexual maturity to the first service an excessive growth is not such a big problem. According to Lammers and Heinrichs (2000) the growth rate over 700 g/day often requires dietary energy concentration to be higher than standard recommended values. Shamay et al. (2005) recommended a moderate feed restriction during the critical period of 3-12 months of age, and added that if there is an undesirable retardation of the calf's growth due to targeted correct development of the mammary gland, maximum skeletal growth and body weight increase by compensation feeding cannot be achieved any more.

Zeman et al. (2006) suggested that young heifers should ideally be given forage only. However, Beck et al. (2005) reported that forage does not supply enough nutrients to meet requirements of intensively reared heifers. In the study by Le Cozler et al. (2009a) effects of grazing of young heifers on their future performance and longevity were investigated. From 4 to 12 months of age the heifers were either at pasture (P), or received maize silage with low nutrient levels (SKN), or with high nutrient levels (SKV). In the period from 4 to 12 months of age, ADG values in different experimental groups were as follows: P = 874 g/day, SKN = 736 g/day and SKV = 890 g/day. The SKV heifers achieved sexual maturity 28 days earlier on average (P<0.001). However, the P and SKN heifers had higher first lactation milk yield and longer production life than the SKV ones. According to Zeman et al. (2006) it is important to check production efficiency of the diet regularly and keep adjusting the diet to meet standard recommended nutrient values (Czech Technical Standards, CSN) . The effects of energy and protein levels in the diet for heifers during puberty also depend on their genetic potential.

2.1.3 Nutrition and growth of heifers during the post-puberty period

The system of nutrition of sexually mature heifers is based on the same principles as cow nutrition systems in terms of production and nutrition management, while respecting their specific requirements for complementary feeds and their physiological requirements including a strict control of nutrient supply to prevent over-conditioning (Zeman et al., 2006; Nor et al., 2013). From the beginning of second year of life there is a risk of over-conditioning. Nutrient concentrations in the diet should be reduced. Practical experience shows that intensively reared heifers that receive diets with adequate proportions of fibre around 12 month of age achieve good reproduction results (Veauthier et al., 2000). Meyer et al. (2004) mentioned that all the dietary mistakes made in the second year of heifer rearing period lead to impaired lactation performance, i.e. reduced daily milk yield and compromised fertility. Roche et al. (2000) found that poor feeding management from 12 to 18 months of age resulted in impaired cow fertility. This indicates that inadequate feeding (energy x protein) can worsen fertility and increase early embryonic death. Shamay et al. (2005) found that a mean increase in body condition score by 0.77 points during 60 days from the onset of puberty is a natural part of sexual maturing process. Bouška et al. (2006) suggested that around the first service a heifer should have BCS of 3. After the breeding at 360 – 400 kg body weight (depending on breed) it is necessary to adjust the diet to weight gain over 800 g/day, and support the body frame growth and suppress fat deposition. Similar body weight gain values (816 g/day) were recommended also by Meyer et al. (2004). St-Piere (2002) recommended the target adult body weight of 630 – 820 kg for Holstein cows. Heifers should achieve 55% of target body weight during the first pregnancy, 85% at the first calving, 92% at the second calving and 96% at the third calving. According to Spiekers et al. (2009) in the older heifers, a large proportion of protein supply is provided by ruminal microflora. Carbohydrate supply in older heifers should be adjusted so to prevent fat deposition. Particularly starch and non-digestible starch should be kept at adequate low levels. Bouška et al. (2006) suggested that from the 7th month of pregnancy the heifer feeding is similar to dry cow feeding, and 3 weeks prior to calving nutrient concentrations in the diet should be increased to levels similar to the production diet which they will receive after the calving, also with respect to adequate mineral and vitamin supply. The ideal body condition score of heifers around calving is 3 – 3,75. Wattiaux (2011) reported that over-conditioned heifers have a higher risk of dystocia and postparturient metabolic disorders. Body condition score is a suitable tool to assess nutrition of pregnant heifers. Patterson et al. (1992) suggested that heifers should receive a well-balanced high energy diet which provides nutrients for adequate growth of the heifer and foetus, without over-conditioning the heifer.

2.2 Assessment of body condition, body weight and other growth parameters in young heifers

In order to determine optimum growth rate of heifers, 3 parameters should be taken into account: optimum age at first calving, optimum height at the croup at first calving and genetic potential for the body frame size (Hoffman, 1997). Mourits et al. (1999a) suggested that to achieve the ideal management it is necessary to systematically evaluate the growth and development of heifers by measuring body weight at certain ages. However, body weight should not be the only criterion because it does not reflect the level of nutrition and body capacity of the heifer (Le Cozler et al., 2008). In order to find eventual deviations from the target growth it is necessary to evaluate skeletal development, too (croup height, pelvic width) and growth of organs, muscles and subcutaneous fat tissue, e.g. by means of body condition scoring (BCS) (Shamay et al., 2005). For the body condition evaluation also a more objective method of ultrasound back fat measurement can be used (Domecq et al., 1995). With this method very good results were obtained for instance in Germany (Schröder and Staufebiel, 2006). The ultrasound examination along with health checks in different growth phases can provide timely information on retarded skeletal growth and/or over- or under-conditioning in the rearing period (Macdonald et al., 2005). BCS is a simple, non-invasive method to assess a current nutritional status of the animal by evaluating fat deposits around the tail head, pelvis and loin (Domecq et al., 1997). Different systems of body condition scoring are described in detail in the review by Roche et al. (2009). Most of them are based on scoring scales to evaluate nutritional status of animals. In the Czech Republic a 5 point BCS scale is used most commonly, with the accuracy of 0.5 or 0.25 points which is based on the original system published by Edmonson et al. (1989).

Figure 1- Body weight (kg), height (cm) and body condition scores (points) in different months of heifer rearing period for small and large breeds (Wattiaux, 2011)

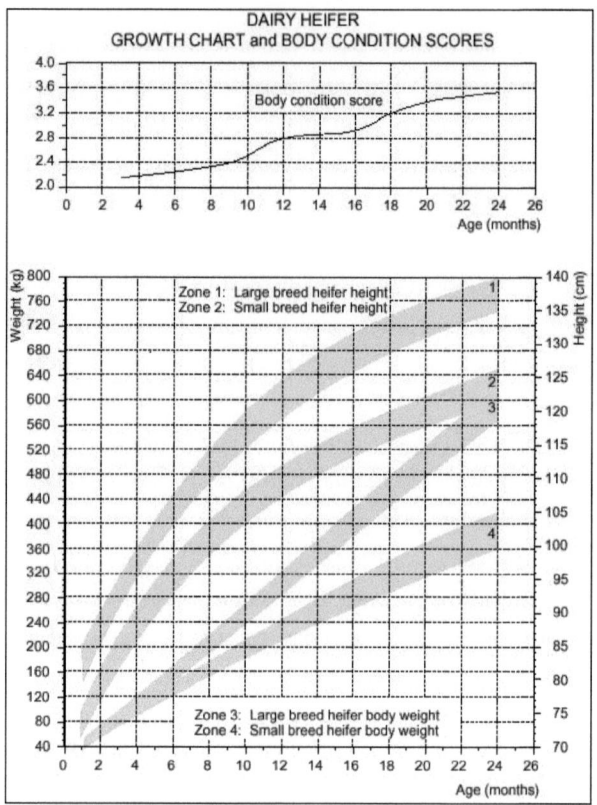

The growth and development of heifers must be compared with the standard growth curve (Figure 1). This requires regular weighing, BCS assessment and wither height measurement in heifers. Such checks should be performed regularly, at short intervals, to be able to adjust nutrition in a timely fashion if necessary (Wattiaux, 2011). Keown (1996) suggested that heifers should be assessed at 6 months of age to check whether they do not grow too quickly or slowly. Thereafter they should be checked at least 3 times before being bred to avoid fertility disorders. It is also important to check heifers 2 months before calving because dietary changes performed in a timely manner may prevent dystocia and post-calving metabolic disorders. Veauthier et al. (2000) pointed out that changes in housing and feeding in different rearing phases may strongly influence the growth curve. Knott et al. (2007) and Svensson et al. (2008) suggested that it is vital to reduce stress to a minimum level, e.g. by maintaining heifer groups.

2.3 Connections between young heifer management and later production performance

Young heifer management strategies depend on breed, genetic potential of animals and target first calving age. The management strategy strongly influences future performance and profitability of the herd (Mourits et al., 1999a). The age at first calving has an impact on heifer-raising cost and consequently on depreciation of production cows, and on the other hand on lifetime performance and longevity of cows. The aim of the heifer rearing is to raise dairy cows that will give maximum performance during their productive life at the lowest possible expense (Stevenson et al., 2008; Le Cozler et al., 2009a).

2.3.1 Relationship between intensity of heifer rearing and later milk yield

Most researchers agree on a negative impact of high average daily weight gain in young heifers on first lactation milk yield. Particularly the pre-puberty period is critical because mammary parenchyma reduction may occur (Capuco et al., 1995; Sejrsen and Purup, 1997; Silva et al., 2002; Ettema and Santos, 2004; Daniels, 2010) although in some experiments no such effects were found (Kerzt et al. 1987; Waldo et al. 1998; Pirlo et al. 2000) and in some mammary parenchyma was reduced only if ADG was over 0.7 kg/day (Abeni et al., 2000), 0.9 kg/day (Knight and Sorensen, 2001), and 1.0 kg/day (Mourits et al., 1999a). Van Amburgh et al. (1998) proposed the upper ADG limit for smooth mammary gland development of 0.7 − 0.8 kg/day. A similar conclusion was drawn by Zanton and Heinrichs (2005) in their meta-analysis of eight studies. Van Amburgh et al. (1998) suggested that both the low (up to 400 g/day) and high (over 800 g/day) ADG before puberty leads to a reduction in milk yield in subsequent lactations by 10 to 40 %. Shamay et al. (2005)

concluded that an ADG of about 0.7 kg/d is optimal for achieving maximum performance. According to Mourits et al. (1999a) a higher intensity of nutrition before sexual maturity impacts the secretion of lactogenic complex hormones, resulting in fewer secretory cells in the mammary gland. Tozer and Heinrichs (2001) and Sakaguchi et al. (2005) suggested that intensive nutrition before puberty has a negative impact in all the breeds, but the type of diet that causes the reduction in milk yield is variable. According to Madgwick et al. (2005) an accelerated growth of Holstein heifers during puberty, enhanced by higher protein levels in the diet, does not depress future milk production. A similar conclusion was drawn also by Macdonald et al. (2005). They suggested that heifers growing faster before puberty show an impaired mammary gland development, but the first lactation milk yield was not adversely influenced thanks to a better body development. Foldager and Sejrsen (1991) reported that by increasing ADG in young heifers from 0.4 to 0.6 kg/day milk yield and mammary gland size were increased by 10 %. A further increase in ADG, up to 0.8 kg/day, did not have an effect on mammary gland volume or milk yield. According to Abeni et al. (2000) ADG of 0.9 kg/day in young heifers significantly reduced milk fat in the first lactation. Hohenboken et al. (1995) stated that the relationship between ADG and milk yield depends on the breed. Small breeds are more sensitive to negative consequences of intensive rearing in terms of the first lactation milk yield (Sejrsen et al., 2000). Heifers with high milk production potential seem to be less sensitive to intensive nutrition. Holstein heifers can achieve higher ADG without marked undesirable effects. Moreover, higher ADG has an accelerating effect on the puberty onset (Le Cozler et al., 2008).

2.3.2 Effect of age at first calving on future milk yield

Economic results of heifer rearing (and, to a certain extent, of cow management, too) are influenced by the length of rearing period. The length of heifer rearing period depends mainly on growth rate and conception at optimum live weight and age. Opinions about an optimum first calving age vary considerably (Mourits et al., 2000). As soon as a heifer reaches body weight considered as ideal for conception, she should be successfully bred. Any prolongation of the empty time leads to mostly undesirable increase in age and body weight at the first conception and calving (Britt et al., 1998; Stevenson et al. 2008).

Mourits et al. (2000) found that with average daily weight gain of 0.9 kg/day before puberty and maximum average daily weight gain of 1.1 kg/day after puberty it is virtually impossible to achieve the first calving at 20.5 months and 563 kg body weight, which leads to a subsequent income of $107 per heifer per year. Some researchers (Britt et al., 1998; Mourits et al. 1999b) did not find any negative consequences of early calving before 22 months of age if the growth phases of young heifers were respected. Ettema and Santos (2004) reported that heifers which calved at earlier ages had lower first lactation milk yield but better lifetime performance. Similar results were reported by Dawson and Carson (2004). They suggested that the first calving live weight of 540 kg is more profitable than 620 kg. Late calving heifers produced by 11 % milk more in the first lactation. Early calving heifers had greater weight losses during early lactation and longer calving-to-calving interval. However, the first calving live weight did not affect milk yield in the second and third lactations (Shamay et al., 2005). Macdonald et al. (2005) suggested that the prepubertal growth rate and body weight at first calving influence milk yield only in the first lactation, but not in the later ones. Shamay et al. (2005) and Wattiaux (2011) confirmed a strong positive correlation between body weight at first calving and first lactation milk yield. Holstein heifers should weigh 620 kg in the first month of lactation to achieve maximum milk yield. According to Hoffman et al. (1996) the growth acceleration along with poor breeding efficiency may lead to negative consequences, i.e. if heifers are fed to calve at 22 months and do not get pregnant till 15 or 16 months, the delay may lead to over-conditioning and increased body weight at first calving. Such heifers are extremely susceptible to ketosis, abomasal displacement and likely to have a low feed intake in the transition period. Mourits et al. (1999a), Heinrichs and Gabler (2003), Ettema and Santos (2004) and Shamay (2005) recommended the first calving age to be 23 to 24 months in Holsteins, to achieve the most profitable milk production. Stevenson et al. (2008) found that only 2.7 % of dairy herds achieved the recommended first calving age of ≤ 24 months, at ≥ 560 kg BW. This leads to economic losses. In the study by Wathes et al. (2008) good fertility results under top performance in the first lactation were achieved with the first calving age of 24 to 25 months. Nevertheless, the heifers that calved at 22-23 months of age showed the best performance and longevity (over 5 years of life), partly because highly fertile heifers keep their good fertility to higher ages. It can be concluded that early calving ages less than 23 months are associated with lower milk yield in the first lactation, lower milk fat, and, on the contrary, higher milk protein, which is also due to a reduced milk production. Moreover, reproductive efficiency is usually lower, too (Van Amburgh et al. 1998; Abeni et al. 2000). Ettema and Santos (2004) reported that the first calving ages higher than 24.5 months did not improve fertility, lactation performance nor the health of heifers. Medium calving (23-24 months) heifers brought by $138.33 and $98.81 higher profit than early calving (before 23 months) and late calving (over 24.5 months) ones, respectively. Hoffman (1997) concluded that mean first calving ages higher than 24 months are not profitable and indicate an uneven herd. Mourits et al. (1999b) suggested to take into account and compare possible benefits of lower first calving ages such as lower feed cost, higher cumulative production per month of age, shorter generation interval and lower overheads, and disadvantages such as reduced conception rate, impaired calving ease, reduced milk yield per lactation, reduced longevity and increased cost of nutrients required in later diets. The following figure adapted from Fricke (2003) summarizes intensities of heifer rearing and optimum age at first calving. According to Fricke (2003) the optimum calving age is 24 months and every day older than 24 months increases the raised heifer cost by $1.50 - $3.00.

Figure 2 – Reproduction management of heifers (Fricke, 2003)

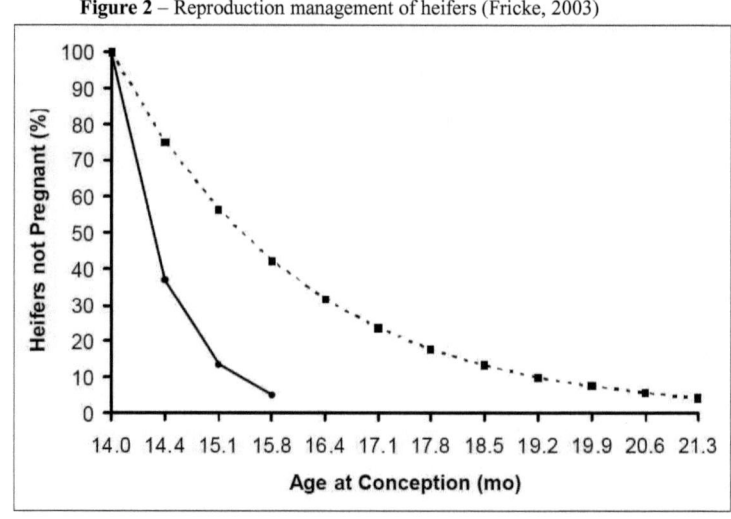

In this scenario, heifers are subjected to oestrus detection at 14 months. The broken line shows the rate at which heifers become pregnant under poor reproductive management (a service rate of 40 percent and a conception rate of 50 percent). By contrast, the solid line shows the rate at which heifers become pregnant under excellent reproductive management (a service rate of 90 percent and conception rate of 70 percent). Although the average age at first calving is 25.4 months for the poor-management group of heifers, more than 25 percent of the heifers will not calve until after 26 months of age and 10 percent of the heifers will not calve until after 28 months of age. Clearly, the average age at calving does not reflect an underlying reproductive problem. For the solid line, average age at calving is 23.9 months. But, more important, 95 percent of heifers subjected to excellent reproductive management will calve before 25 months of age (Adapted from Fricke, 2003).

As previously suggested, the calculation of economic benefits of early calving is based on heifer-raising cost and lifetime performance, which can be misleading. These arguments are used to demonstrate the benefits of early calving and do not reflect a potential effect of calving age on a herd culling rate (Hoffman et al., 1996). Annual herd replacement cost is also increased when the first with calving age is 20 months because there are some problems associated with early calving. In heifers calving at extremely young ages important biological mechanisms (reproduction, milk synthesis, parturition, feed intake and health) may fail dramatically. Reasons for such disorders are disputable, however, all the shortcomings result in increased herd culling rate (Spiekers and Potthast, 2004). Bouška et al. (2006) reported that, theoretically, herd replacement costs are lower with the calving age of 22 months. For every month the first calving age is higher than 22 months the production cost is increased by about 1,800 CZK per animal due to reduced milk sales and fewer productive days. Good management can help maintain post-calving body weight over 550 kg under conditions that allow many groups of heifers to calve at 22 months of age. According to Kvapilík (2010) the economic loss incurred due to prolongation of the heifer rearing period by one month above the optimum first calving age (usually 23 to 26 months) is commonly reported as 700 to 1800 CZK per heifer. An occurrence of diseases in heifers and calf losses play an important role in cost effectiveness. Kvapilík (2010) estimated the losses of approximately 2,500 CZK for stillbirth and up to 9,000 CZK for calf dying at 6 months of age. The above figures also indicate an additional income for extra milk produced by better reared and earlier calving heifers. This can be expressed as income over feed cost (IOFC) for every extra litre of milk sold, which ranges between 3.8 and 4.8 CZK. With the above-mentioned increase in the first lactation milk yield by about 450 kg milk this means by 1,800 CZK higher net return for the 1st lactation. A similar principle can be used to increase lifetime production of better raised heifers (Vacek et al. 2013). Also reproduction and veterinary costs can be expected to decrease. The improvement of longevity can be expressed as replacement cost reduction, so called cow depreciation which along with a reduction in annual culling rate from 35% to 30% represents a decrease in costs per 1 litre of milk by 0.13 CZK (Kvapilík, 2010).

Hoffman (1997) reported that different heifer growth rates among herds were due to environmental factors (parasites, diseases, ventilation, contamination) rather than nutrition. Therefore Drevjany et al. (2004) emphasized the importance of growth monitoring to avoid too slow or too intensive and consequently too expensive heifer rearing. Kadokaw (2006) emphasized that the herd management practices should be based on the fact that animals have certain biological limits that are important as a "reserve" for their heath and reproduction.

2.4 Profitability and economic parameters of dairy production

The target of every business is to achieve profit. Profit is a difference between total sales (return) of market products and costs of their production. Profitability is the ratio between profit and certain basis, e.g. total assets or net assets, sales, product price/cost (Synek et al., 2011). Therefore the main preconditions for reliable calculations of profit and other economic parameters are knowledge of the volume of sales (total return), cost of the evaluated commodity, period of time, accuracy of measurement, and adherence to the same methodology when comparing results in time series (Kvapilík, 2010). Profit and profitability are crucial for making strategic and tactical business decisions; in short-term (operational) decisions these parameters can fade into the background (Synek et al., 2011). According to the study by Widiati et al. (2012) the profitability of a business can be efficiently evaluated according to breakeven which is defined as such volume of production (sales) when sales are equal to costs but profit is not achieved yet. According to Kvapilík et al. (2012) the main factors that can improve economic results of milk production include milk adequate to production conditions, good health status of cows and related good fertility, adequate herd replacement rate, low mortality and low culling rate, high lifetime production of cows (longevity), high quality forage, balanced diets, high quality of market production, reliable stockpeople, good management and work organization and maximum amount of received subsidies.

Mourits et al. (1999a) and Shamay et al. (2005) reported that heifer-raising cost constitutes about 15 – 20% of total milk production costs.

2.4.1 Structure and calculation of costs

The methodology according to Poláčková et al. (2010) that is recommended for agricultural businesses, includes the calculation equation which contains the following items: purchased feed and bedding, own feed and bedding, drugs, disinfectants, other direct material, other direct costs and services, total labour costs, depreciation of fixed intangible and tangible assets, depreciation of adult animals (previously basic herd), costs of ancillary activities, administrative expenses and total costs. According to Synek et al. (2011), for the management of costs it is necessary to follow them also from a factual point of view that is according to outputs (products and services), i.e. calculation of own costs. Their importance is multiple: in the enterprise they serve to determine internal prices of outputs, preparation of budgets, control and analysis of production economy and output profitability, cost limiting, etc.

Due to a specific nature of livestock production determined by its biological basis and rearing technologies, the calculation of total cost in livestock production is rather complicated. The purpose of calculation is to express expenses both per animal in each production phase and per non-living products (i.e. milk). Because the intensity of product movement (intermediate products) between different animal categories is high, it is necessary to follow these moves in terms of cost and nature. In the interlinked production chain the different stages of intermediate product are gradually associated with external costs and transferred to subsequent production steps (Poláčková et al., 2010). According to a traditional method of calculation of complete (own) costs the principle is based on ascertaining of all costs spent on calculation unit of particular product or service. For the calculation of their total costs agricultural enterprises use different procedures because there are no regulations to define obligatory procedures (Kvapilík a Syrůček, 2012). The cost calculation in livestock production determines that it is necessary to perform a two-step cost calculation. In the first step costs of the main product(s) are calculated. From total costs of animal rearing (production category) the cost of side products is subtracted. In the second step cost per 1 kg body weight is calculated. Here animal movements (intermediate product) among different production categories must be considered. The aim of calculation is to ascertain total expenses per animal and total body weight of an animal in a given phase of rearing or to date The calculation of costs, revenues (sales), profit and other economic parameters in agriculture is usually divided into calculation of total (own) costs and partial costs, or assistance towards costs. While in the Czech Republic and some other EU member states (mainly new ones) the main method is calculation of total (own) costs, in the EU-15 countries the calculation of assistance towards costs prevails (Kvapilík a Syrůček, 2012).

2.4.2 Management and dynamics of dairy herd replacement

Dairy herd is a complex system with two interconnected parts: cow herd (including dry cows) and replacement heifer herd. An intensity of mutual interactions is influenced by a number of cows to be culled and, reversely, a number of heifers needed for herd replacement (Kristensen, 1992). This mutual influence is necessary for understanding the dynamics of dairy herd and subsequent economic impact of replacement heifer rearing (Tozer and

Heinrichs, 2001; Heikkila et al., 2008). Probability of good return on investment is dramatically improved if the herd management is based on good heifer rearing. Good heifers are those that successfully calve and start their productive period (Ettema and Santos, 2004; Bach and Ahedo., 2008). As a starting point the rearing intensity should be considered in the context of herd possibilities and then the goals should be set and a system suitable for achieving them. The economic evaluation of benefits from lower first calving ages is considered as important (Mourits et al., 1999a; Hoffman, 2009).

The level of reproductive performance directly influences the economics of milk production and herd replacement. In recent years fertility in high-producing cows has deteriorated leading to lower profitability (De Vries a Risco, 2005; Leroy a Kruif, 2006). According to Leroy and Kruif (2006) and Lee and Kim (2007) poor heat detection is a primary reason for low conception rate in the herd. A successful reproduction program in the dairy herd increases the probability of longer stays of cows in the herd because it minimizes culling due to poor fertility, increases a proportion of cow life spent in the productive parts of the cycle and also increases a number of heifers available for sale. A well-managed herd can achieve the rate of culling due to poor fertility lower than 8 to 10%. Lucy et al. (2001) reported that an increase in production often leads to worsening of reproductive performance. This is a common situation although some publications claim that it is due to inability of adjust conditions (particularly nutrition) to the requirement of the animal. Heikkilla et al. (2008) found that it is impossible to determine optimum herd turnover that would apply to any herd because there are some other factors in play such as different cow production capacities and their different frequencies of occurrence in the herd. Přibyl (1997) reported that a higher herd replacement rate based on earlier calving heifers may also lead to a greater increase in genetic merit and thereby improvement of herd performance and economic efficiency. On the contrary, Heikkila et al. (2008) stated that a short production life leads to higher replacement cost and reduced capacity of selection and genetic improvement in the herd. Also Honarvar et al. reported that the prolongation of production life resulted in increased profitability. On the other hand, the prolongation of production life lead to a decrease in breeding value for kg milk from 101.24 kg to 87.56 kg within one year. Nevertheless, heifer raising cost was reduced and total sales increased due to sales of surplus heifers.

Heikkila et al. (2008) reported that before culling a cow it is necessary to consider the importance her genetic potential and therapeutic expense, at least in cows with high production capacity, rather than replace her right away because she is ill. According to Pryce et al. (2001) the cows with a higher genetic potential for milk production have usually longer calving interval and better lactation persistency. Kadokawy (2006) reported that a prolonged calving interval may help animals with extremely high production which do not always benefit from high milk yield and shorter calving interval. On the other hand, a prolongation of calving interval means fewer calves born and reduced average daily milk yield in the herd, which leads to a loss of milk sales the magnitude of which grows with an increasing number of days above the optimum calving interval (Kvapilík, 2012). Hulsen (2011) concluded that for high-producing cows the housing technology is very important (floor surfaces, heat stress, overcrowding – increased cow density, e.g. in rooms where cows wait before entering the milking parlour). The quality of management depends on whether routine and structured observations of cows are performed. Stress causes reduced feed intake and immune suppression. Cows feel much safer if they have enough feed, know where escape paths are and can rely on predictable and easy behaviour of stockpeople. Hulsen and Swormink (2006) stated that it is vital for good herd management to check cow health status regularly and intervene immediately if disease signs are observed.

2.4.2.1 Economic parameters of dairy herd replacement dynamics

Dairy producers strive to find such dairy herd replacement policies that would optimize net current cow value, i.e. current and future net profit from an individual cow and all the other cows that will replace her in future (Heikkila et al.,2008). The computation to calculate optimum herd replacement rate is based on a production function (Dijkhuizen, 1992) and depends on a curve of marginal net income (i.e. net income in each year, month or day of life), on traits of a given animal, discount rate and on a real culling date. Therefore marginal net income must be used in the planning horizon. Current net income can be calculated along with expected costs and revenues by means of discounting future costs and revenues (the discounting is a mathematical procedure which discounts, i.e. re-calculates and adds up, future revenues in different periods to a current investment value, using discount rate, i.e. estimated rate of return). If discounting is used then the optimum culling time comes when marginal net cow return becomes equal to maximum annual return on a replacement heifer. A higher discount rate might lead to either later or earlier culling of a cow, depending on the shape of marginal net return curve (Dijkhuizen a Morris, 1997). The economic model by Heikkila et al. (2008) recommends to cull the oldest cows when milk prices drop or prices of reared heifers increase.

The example calculation bellow is based on average present performance of an animal in the herd, which is the most suitable estimated parameter to be compared with expected future returns on replacement heifers arriving in the herd. Future costs and returns are weighted by probability of survival of the animal:

$$\text{ANR}_j = \left(\sum\nolimits_{i=1..j} pi \times MNR_i \right) \bigg/ \left(\sum\nolimits_{i=1..j} pi \times 1_i \right) \qquad (1)$$

ANR$_j$ expected annual average net return;

i   moment of decision on culling or keeping a cow in the herd (1≤i≤j) that comes in the end of the i period;

j   last period when a cow can be culled;

p$_i$ probability of survival of a cow till the end of the i period, considered from the beginning of first lactation (end of the 0 period);

l$_i$ length of the i period (years)

MNR$_i$ marginal net return in the i period including a correction for carcass value change and financial loss associated with culling (Dijkhuizen and Morris, 1997).

If optimum "service life" is determined, Retention Pay Off (RPO) or loss can be estimated by model simulation of the impact of keeping a cow in the herd over the optimum time as compared with an instant culling. The calculations should include risks ensuing from premature culling of cows. RPO is calculated as follows:

$$\text{RPO}_i = \sum\nolimits_{j=i+1..r} p_j \left(MNR_j + ANR_{max} \times 1_j \right) \qquad (2)$$

RPO$_i$ "Retention Pay Off" at the decision moment i;

r   optimum culling moment;

p$_j$ probability of survival till the end of the j period, calculated since the decision moment i;

l$_j$ length of the i period (years);

MNR$_j$ marginal net return in the j period

ANR$_{max}$ expected maximum annual net return (Dijkhuizen and Morris, 1997).

RPO is an economic index that enables to rank animals according to their future performance or profitability, i.e. the higher the RPO value the higher the animal value. Negative RPO values imply that culling is the most profitable decision in a given case (Kalantari a Cabrera, 2012).

Dynamic programming (DP) is a mathematical method that can be used in situations when a sequence of separated decisions is to be evaluated, like in the dairy herd management. DP is based on repeatability of separated decisions to shorten the computing time. DP uses seemingly simple but efficient principles based on the Bellman´s Principle of Optimality (Bellman, 1957).

2.4.2.1.1 Markov chains

DP based on the Bellman´s Principle of Optimality is still the best available technique to calculate economic value of a dairy cow (De Vries, 2006; Cabrera, 2010; Kalantari et al., 2010). However, it is a high complexity of the model that is a drawback of DP. The model gets too extensive and complicated easily (Demeter et al., 2011), with limited usability for practical dairy herd management, and therefore it is not suitable for daily use on farms (Groenendaal et al., 2004). For this reason, Groenendaal et al. (2004) decided to use only the method of marginal net return calculation, without calculating complete DP, and found that this method provided a more accurate model of herd replacement and the results were in agreement with previous studies that used a more complex DP.

Dairy cow economic value of cow is a difference between discounted net return of the cow evaluated and average value of a culled cow (Eicker a Fetrow, 2003). If the value is positive, it is recommended to keep the cow in the herd. If the value is negative, the cow should be culled. Such decisions when implemented would result in improved total net return of the whole herd (De Vries, 2004). Standard DP is an iterative process of optimisation which always uses a reverse approach: from the future back to the past. Every iteration is based on a value of cow either kept in herd or culled. Standard DP uses as many such values as possible to connect different iterations that subsequently enable to achieve the optimum herd replacement rate (Cabrera, 2012). Important factors that influence the result include lactation phase, pregnancy, milk yield, heifer genetic merit and others (De Vries 2004, 2006). DP can create a matrix of cow economic values for every stage of her life. Nevertheless, other calculations within DP can be simplified or skipped when using the Markov chain model for the calculation of whole herd total net return including the herd structure and other economic parameters (De Vries 2004, 2006; Kalantari et al., 2010). Markov chains is a simulation technique that requires pre-defined and if possible true values of reproductive performance and herd replacement rate. This interactive process enables to consider also other parameters of reproduction such as anticipated target, and subsequently compare the changes of management interventions in the herd (Cabrera, 2012). Markov chains contain a definite number of possible stages, their classification and a limited number of time parameters in the matrix X (matrix for different life stages of an animal in the herd). In the transitional matrix or vector P transitional probabilities between the stages are expressed. The individual stages can be repeated, transitional or immersive. After a certain period of time the transition matrix P (iteration process) comes to a state of equilibrium which is independent of the original distribution of probabilities and from that point on the program calculations involve the structure of the whole production system in its stable state (Dijkhuizen a Morris, 1997).

Figure 3 – The use of Markov chains to calculate dairy cow economic value (matrix x contains: MIM – month in milk; PAR - parity and MIP – 0 for pregnancy and 1 to 9 for different months of pregnancy) (Cabrera, 2012)

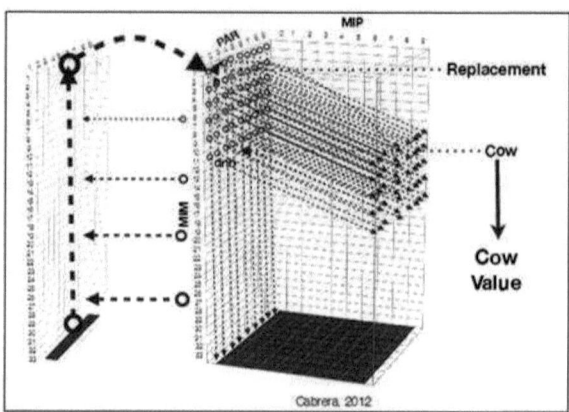

In the study by Cabrera (2012), see Figure 3, the herd structure is calculated on a monthly basis, using the Markov chains model, and transitional vector P between different stages contains probabilities for culling, pregnancy and abortion, i.e. probabilities of leaving the herd for the cow (change of stage).

21

By means of Markov chains the simulation of herd dynamics is performed. The program considers the structure of the whole production system in its stable state. Economic efficiency of the system is expressed as a function of biological context, herd management and economic parameters (Cabrera, 2012).

3 Hypothesis and objective

- working hypothesis
- It can be assumed that on the basis of knowledge of biological connections between heifer rearing intensity and performance and longevity of cows in the production period it is possible to evaluate economic efficiency of heifer rearing techniques with a view to economics of dairy production.

- objective
- The aim of the study was to define connections between heifer rearing intensity (i.e. heifer growth represented by body weight and body condition score in the period under study), first calving age and subsequent performance and longevity of dairy cows in the production stage.

4 Material and methods

In several selected Holstein (H) and Czech Fleckvieh (FV) dairy herds in the Czech Republic production and economic parameters of heifer rearing were analysed with regard to subsequent production and economic results of dairy production. The database with heifer rearing data was evaluated using software developed by ourselves within SAS 9.2 (2008) and system of general procedures supported by MS Excel and R-software.

Connections between heifer rearing intensity and parameters of production period were evaluated using linear mixed models, specifically the procedures MIXED (Verbeke and Molenberghs, 2000), REG, CORR and LOGISTIK in the statistic software SAS 9.2 (2008). Taking into account methods used by other researchers (Gabler et al., 2000; Pirlo et al., 2000; Hultgren et al., 2009a, 2009b, 2010), and explained variation measure and statistical significance of computing equations, the most suitable model of data evaluation was chosen.

4.1 Characteristics of data files and method of evaluation

The source data for the evaluation of management and economic parameters of heifer rearing were provided by two databases created (see Chapters 4.1.1 and 4.1.2). The data were obtained jointly with the farm ŠZP Lány (owned by Czech University of Life Sciences Prague) and VÚŽV v.v.i (Institute of Animal Science) Praha-Uhříněves.

4.1.1 Database for the evaluation of heifer rearing and cow production

The source data for the evaluation of heifer rearing and the subsequent production and reproductive performance were obtained from 2 farms (ŠZP Lány – Czech University of Life Sciences Prague, and Netluky Prague-Uhříněves) between 2005 and 2012. The source file of farm records contained data on body condition score, body weight and daily weight gain at monthly intervals from 5 to 18 months of age from 780 Holstein heifers. The data file also contained data from the first three lactations of the raised heifers (official milk test results): milk yield (kg per lactation), milk protein content (%, kg), milk fat content (%, kg), calving to first service interval (days), days open (days), member of services per conception (index), calving interval (days), standardized 100 day milk (kg per lactation), 100 day milk protein

24

content (%, kg), 100 day milk fat content (%, kg), first calving age (days), calving year and month, and sire breeding value for kg milk.

Heifers were reared under similar conditions and weaned approximately at 3 months of age. Total mixed ration (TMR) was given as two different diets (TMR1: heifers up to 12 months of age, ME to PDI ratio 39-66:1 g/Mcal; TMR2: from 12 months of age, ME to PDI ratio 40-45:1 g/Mcal). The heifers were bred when they reached body weight of 400 kg ≈ 14 months of age. Mean lifetime production (**PCU**) (milk yield, milk protein % and milk fat %) was calculated as mean milk yield for the first three concluded lactations. At the beginning of data collection the animals were not of same age, therefore the evaluated data are not uniform (see Tables 1 and 2). As to the evaluation of milk yield, only cows with more than 250 days in milk (DIM) were included. Afterwards, lactation yield was extrapolated and standardized for a 305 day lactation. For the first lactation also first 100 DIM were evaluated to assess the effect of growth and body condition of heifers on this period.

Table 1 summarizes the evaluated parameters of heifer rearing including basic descriptive characteristics. Table 2 gives basic descriptive characteristics of production and reproductive parameters under study during the first three lactations (dependent variables). The Figure 4 shows first calving age histogram of the heifer population under study.

Table 1 – Basic characteristics of growth parameters during the heifer rearing period

Parameter	n	$\bar{x}$	sd	min	max
Body condition at 14 months (points)	530	3.37	0.33	2.00	4.50
BW at 14 months (kg)	780	412.49	37.50	250	530
ADG from 5 to 10 months of age (kg)	392	0.91	0.11	0.54	1.21
ADG from 11 to 14 months of age (kg)	528	0.91	0.08	0.62	1.17
ADG from 5 to 14 months of age (kg)	370	0.91	0.09	0.58	1.19
AFC (days)	780	727	58	581	1 051

Table 2 – Basic characteristics of the first three lactations in the evaluated heifers

Parameter	n	$\bar{x}$	sd	Min	Max
Services per conception	780	1.95	1.34	1	9
1st lactation					
Milk yield (kg/ 100 DIM)	780	2 916	507	852	4 595
Protein % (100 DIM)	780	3.91	0.53	2.42	4.46
Fat % (100 DIM)	780	3.09	0.19	2.55	3.74
Milk yield (kg/305 DIM)	780	8 511	1 296	3 624	13 475
Protein % (305 DIM)	780	3.25	0.17	2.72	3.78
Fat % (305 DIM)	780	3.85	0.44	2.67	5.28
Services per conception	428	2.35	1.70	1	10
Open days	702	149	85.24	39	525
Calving interval (days)	500	415	83.28	267	801
2nd lactation					
Milk yield (kg//305 DIM)	443	10 116	1 779	5 042	16 877
Protein % (305 DIM)	443	3.24	0.20	2.77	4.06
Fat % (305 DIM)	443	3.65	0.43	2.41	5.07
Services per conception	283	2.24	1.52	1	9
Open days	367	157	84.20	42	482
Calving interval (days)	232	415	79.95	291	803
3rd lactation					
Milk yield (kg//305 DIM)	194	10 443	1684.19	5 916	16 749
Protein % (305 DIM)	194	3.17	0.17	2.78	3.47
Fat % (305 DIM)	194	3.60	0.41	2.45	4.60
Open days	194	153.17	76.96	38	437
Mean lifetime performance (PCU)					
Milk yield (kg/305 DIM)	194	10 032	1 221	6 481	12 958
Protein % (305 DIM)	194	3.20	0.16	2.82	3.78
Fat % (305 DIM)	194	3.65	0.39	2.75	4.86

Figure 4 – Histogram of age at first calving distribution (AFC)

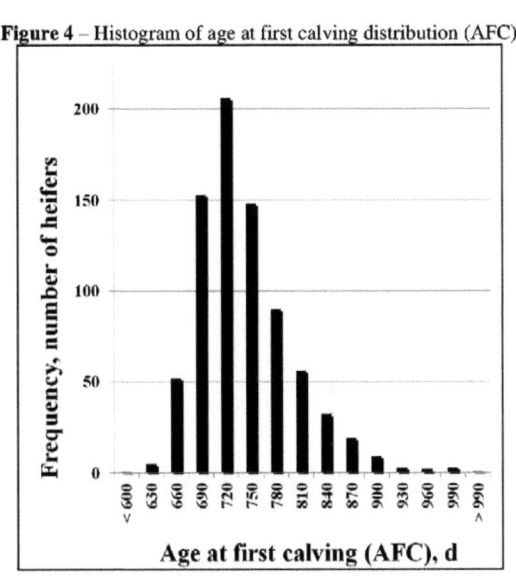

4.1.1.1 Process of evaluation of heifer growth and production period

Average daily weight gain (ADG) of heifers was divided into three periods (Table 1): prepubertal growth (5 – 10 months), postpubertal growth, i.e. from the onset of sexual maturity to breeding (11 – 14 months) and total growth (5 – 14 months). All the explaining variables given in Table 1 (body condition score and body weight at 14 months, ADG 5 – 10 months, ADG 11 – 14 months, ADG 5 – 14 months, and first calving age) were divided into groups listed in Tables 5 and 6 in Chapter 5 along with counts of animals in the groups, and evaluated.

The Figure 4 shows that the age at first calving most frequently ranges between 23 and 24 months, i.e. age at first service 14 to 15 months, therefore 14 months of heifer age was considered as the first service age and in the part containing the evaluated variables it is the bordeline month for the rearing period evaluation (Table 1).

For the evaluation of heifer growing period and following production period, the equations 1 and 2 used the MIXED SAS 9.2 (2008) procedure. To compare the means, the Tukey method was used (Verbeke et al., 2000). The models contained fixed effects only and individuals were considered as independent observations.

$$y_{ijklmn} = \mu + A_i + S_j + H_k + B_l + BV_m + b(age_{ijklm} - age_{00000}) + e_{ijklmn} \qquad [1]$$

y_{ijklmn} = dependent variables (listed in Table 2), μ = mean, A_i = effect of calving year (i = 2007 (n=125), 2008 (n=154), 2009 (n=164), 2010 (n=230), 2011 (n=107)), S_j = effect of calving season (j = spring (n=213), summer (n=231), autumn (n=196), winter (n=140)), H_k = effect of herd , B_l = independent (explained) variables (effect of body condition or body weight or average daily weight gain categories, given in Tables 1, 5 and 6), BV_m = sire breeding value for milk yield (kg) (m = $\geq$750 (n=260), 749-300 (n=257), $\leq$299 (n=263)), b = vector of regression coefficients of first calving age used for B_l, but only for analysis of production and reproduction parameters in the first three lactations, age_{ijklm} = age at first calving in days, age_{00000} = average age at first calving, and e_{ijklmn} = random error.

The simplified equation 2 was used to evaluate the chosen variables which were ascertained during the heifer rearing phase.

27

$$y_{ijklm} = \mu + C_i + D_j + H_k + B_l + e_{ijklm} \qquad [2]$$

y_{ijklm} = dependent variables: i.e. body condition score at 14 months of age, body weight at 14 months of age, number of services per conception in heifers and first calving age, μ = mean, C_i = effect of year of birth (i = 2005 (n=138), 2006 (n=136), 2007 (n=154), 2008 (n=245), 2009 (n=107)), D_j = effect of season of birth (j = spring (n=183), summer (n=195), autumn (n=210), winter (n=192)), and e_{ijklm} = random error.

Differences between the variables were tested at a significance level $P < 0.05$.

To evaluate first lactation culling rate the procedure LOGISTIK was used, Figure 6, Chapter 5.1.1.

4.1.2 Database for the evaluation of economic parameters

Data needed for the analysis of economic parameters are based on data obtained from 33 dairy farms in 2011. Specifically, these were average values for the year of 2011 from 17 Holstein herds, 8 Czech Fleckvieh herds and 8 mixed (H + FV) herds. The data were collected by means of questionnaires (Annex 1). In order to get more information on data file, average costs and standard deviations of heifer rearing are provided that were calculated for a day on feed, raised heifer and kg of weight gain (Table 3) according to a certified method (Poláčková et al., 2010). Table 4 shows average values of the most important management and economic parameters of the Czech herds under study. The evaluated variables (Tables 3 and 4) are designated with *. The investigations done by the Institute of Animal Science (VÚŽV) included, in addition to management and economic parameters of heifer rearing, a questionnaire covering the whole dairy production system. Figure 5 shows the cost structure of dairy production on the 33 farms included in the study.

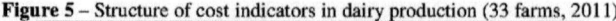

Figure 5 – Structure of cost indicators in dairy production (33 farms, 2011)

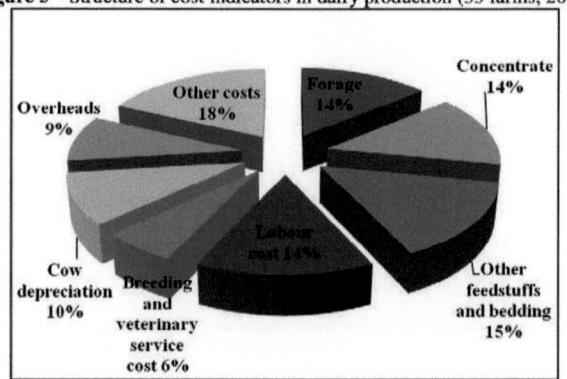

The Figure 5 shows that the greatest part of total costs is feed cost, labour cost and other costs. Nevertheless, the item "cow depreciation", i.e. the difference between price (cost) of fresh heifers and slaughter price of culled cows is also relatively high, and there is also a great variability between herds (CV=61.35%), therefore a potential to reduce costs associated with the herd replacement.

In addition to real data, subjective farmer views of effects of some factors on heifer rearing success were evaluated (Figure 10) in Chapter 5.2.1.

Table 3 – Average heifer raising costs (33 farms, 2011)

Parameter	CZK per						%
	DOF		ADG (kg)		Heifer		
	$\bar{x}$	sd	$\bar{x}$	sd	$\bar{x}$	sd	
Number of farms evaluated	33						
*Total feed cost (summarized[1])	21.71	-	30.31	-	10 124	-	51.44
Milk + milk replacers[1]	3.88	3.09	5.43	4.14	1784	1415.77	9.06
Forages[1]	11.80	4.63	16.56	8.98	5512	2747.31	28.01
Concentrates[1]	6.03	3.33	8.32	5.41	2828	1937.23	14.37
Labour cost	5.39	2.95	7.52	4.74	2585	1934.71	13.44
POL(Petroleum, Oil and Lubricants) and energy	1.29	1.00	1.80	1.73	608	557.52	3.09
Total vet. and herd improvement costs (summarized[2])	2.36	-	2.32	-	901	-	4.58
of this – vet. services and drugs[2]	0.96	0.71	1.35	1.14	453	356.51	2.30
of this – insemination and breeding services[2]	1.01	0.69	1.36	1.04	448	299.15	2.28
Depreciation of tangible fixed assets	1.09	0.68	1.45	1.00	541	419.35	2.75
Overheads	6.59	5.60	9.38	9.16	3116	2630.20	15.83
Other costs	4.15	4.37	5.70	6.23	1805	1662.93	9.17
*Costs per heifer[3] (average)	48.74	21.61	68.10	35.48	22660	11914	-
Costs per heifer (sum of averages)	42.19	-	58.87	-	19680	-	100
*Costs per calf at 6 months of age[4]	53.24	19.02	64.56	22.73	13 144	5946.71	66.79

[3]Weight gain 0.73 g/head/day; [4]Weight gain 0.83 g/head/day, body weight 193 kg at 6 months; *evaluated variables

The data show (Table 3) that heifer raising costs reached 22,660 CZK on average. The largest expense is made up by feed costs (51.44%), overheads (15.83%) and labour costs (13.44%). Calf raising costs up to 6 months of age amounted to approximately 67% of total costs. It should be noted that there was a great variability of costs in the data set.

Table 4 – Average values of selected management and economic parameters (33 farms, 2011)

Parameter	n	$\bar{x}$	sd
No. of cows per herd	33	697.21	413.30
*Milk (kg)	33	7 989.03	1 324.99
*Milk protein %	33	3.40	0.13
*Milk fat %	33	3.87	0.21
*Heifer ADG (kg/head/day)	33	0.73	0.15
Annual milk sales (CZK/cow)	33	62 854.89	10 852.16
Annual milk sales (CZK/DOF)	33	171.94	30.26
Milk price (CZK/L)	33	8.30	0.33
*Cow depreciation (CZK/L of milk)	33	1.54	4.09
*Cow depreciation (CZK/cow)	33	11 140.79	29 120.99
*Cow depreciation (CZK/DOF)	33	30.47	79.66
*Total profit, without subsidies (CZK/L milk)	33	-0.62	1.76
* Total profit, without subsidies (CZK/cow)	33	-4 074.47	12 440.84
* Total profit, without subsidies (CZK/DOF)	33	-11.20	33.88
* Profitability (%)	33	-2.38	27.17
*Conception rate after 1st insemination (%) (heifers)	33	60.51	7.94
*Conception rate after all inseminations (%) (heifers)	33	60.21	11.27
* Conception rate after 1st insemination (%) (cows)	33	39.87	8.37
* Conception rate after all inseminations (%) (cows)	33	42.10	11.72
*Cow culling rate (%)	33	34.78	10.43
*Services per conception	33	2.22	0.35
*Open days	33	120.74	20.48
*Calving interval (days)	33	405.77	17.63
*AFC (days)	33	777.95	37.50
*Number of lactations	33	2.40	0.43
Number of calves raised per 100 cows (heads)	33	100.30	21.43
*Calf death rate (%)	33	5.08	3.71
*Total calf losses (%)	33	11.62	4.74

evaluated variables

Production, reproduction and economic parameters were monitored in a group of mothers of heifers under study. The heifer mothers were raised in the same system as their daughters and therefore their average first calving ages and average daily weight gain can be predictive for those expected in their daughters. In spite of a higher breeding value of heifers, only negligible changes in management and conditions on the farms were found in relation to older cows replaced by the heifers.

Costs of milk production were calculated according to a certified methodology (Poláčková et al., 2010).

The profit excluding subsidies was calculated as follows:

$$CN2 = CN1 - VN,$$ [3]

$$CZ = CZM - CN2,$$ [4]

CN1 = total cumulated costs /year, VN = side product costs /year (calves, manure), CN2 = total cumulated herd costs without side product costs /year, CZ = total profit excluding subsidies /year, CZM = total profit from milk sales /year.

The equation for CN1 calculation includes costs of purchased feed and bedding, farm´s own feed and bedding, drugs and disinfectants, other direct costs and services, labour costs, depreciation of fixed intangible and tangible assets, depreciation of mature animals, expenses on ancillary activities and overheads (Poláčková et al., 2010).

Profitability of costs (R, %) is the ratio of profit and a basis with which the profit was achieved such as assets, costs, sales, performances etc. Profitability is included in the basic financial analysis of an enterprise (Synek et al., 2011). In this study (and in agriculture in general), profitability of costs is commonly used, i.e. the ratio between economic result (profit or loss) and total cost incurred. Profitability determines economic benefit achieved by a company by spending certain means (costs).It is given in percentages per accounting period (usually one year) and the main advantage is the possibility to perform year-to-year comparison among enterprises within the industry, regardless of size (e.g. herd size). Unlike traditional profit criterion (absolute profit), the profitability takes into account also expenses that had to be spent to achieve profit. The basic requirement is that this parameter be a positive value (Poláčková et al., 2010).

$$R = \left(\frac{CZ}{CN2} \right) \times 100.$$ [5]

4.1.2.1 Evaluation procedure of economic parameters

First calving age, average daily weight gain and milk yield were the evaluated variables. Chapter 5 contains Tables 7 and 8 with results. For the evaluation itself the MIXED model SAS 9.2 (2008) was used. Means were compared by the Tukey method (Verbeke and Molenberghs, 2000).

$$y_{ijkl} = \mu + B_i + R_j + D_k + e_{ijkl},$$ [6]

$\mathbf{y}_{ijkl}$ = dependent variables (listed in Tables 3 and 4 and designated with *); μ = mean; $\mathbf{B}_i$ = effect of breed (i = 17 Holstein herds, 8 Czech Fleckvieh herds, 8 mixed Holstein and Fleckvieh herds); $\mathbf{R}_j$ = effect of farm region (j = number of farms [out of total 33 farms] in different regions: South Bohemia – 1; South Moravia – 3; Hradec Králové – 2; Liberec – 1; Moravian-Silesian region – 2; Olomouc – 3; Pardubice – 4; Pilsen – 4; Central Bohemia – 6; Ústí nad Labem – 2; Bohemian-Moravian Highlands – 4; Zlín – 1); $\mathbf{D}_k$ = effect of first calving age, average daily weight gain or milk yield (listed in Table 4) were divided into groups (Tables 7 and 8); and $\mathbf{e}_{ijkl}$ = random error. Breed ($\mathbf{B}_i$) and effects $\mathbf{D}_k$ were considered as fixed effect and farm region ($\mathbf{R}_j$) as random effect.

Differences between the variables were tested at a significance level of $P < 0.05$.

Regression coefficients and correlations were calculated to better understand relationships between selected factors (mean farm values): first calving age and average daily weight gain, first calving age and calving interval, first calving age and open interval, average daily weight gain in heifers and cow depreciation (CZK/head/year), milk yield/year, and profit/cow/year, milk yield/year and average daily weight gain in heifers. For the calculation of regression coefficients the same model was used (equation 6), but only with the inclusion of breed effect ($\mathbf{B}_i$) and farm region effect ($\mathbf{R}_j$). The above-mentioned regression and correlation coefficients are mentioned in Chapter 5.

5 Results

5.1 Evaluation of biological context of heifer growth

A relationship between the heifer growth and reproductive and production parameters later in their life was evaluated to understand optimum dairy herd management principles. The Chapter 4.1.1. gives the data used and calculation procedure. The aim of this dissertation was to analyse values of body age at first calving **(AFC)** in Holstein heifers during the rearing period and their effect on performance during the three following lactations. AFC depends on heifer rearing intensity and producers aim to shorten this period as much as possible. In this study no negative effect of AFC under 23 months was found. Tables 5 and 6 and Figures 6, 7 and 8 give the results we found which are described in the following chapters.

Table 5 - Effects of heifers' average daily weight gain (ADG) on production and reproduction outcomes

Item	ADG 5 – 10 mo of age (≥0.970)	ADG 5 – 10 mo of age (0.969 - 0.850)	ADG 5 – 10 mo of age (≤0.849)	ADG 11 – 14 mo of age (≥0.950)	ADG 11 – 14 mo of age (0.949 - 0.850)	ADG 11 – 14 mo of age (≤0.849)	ADG 5 – 14 mo of age (≥0.950)	ADG 5 – 14 mo of age (0.949 - 0.850)	ADG 5 – 14 mo of age (≤0.849)
n	122	167	103	162	239	127	119	156	95
BCS 14 mo	3.43±0.03 [a]	3.26±0.03 [b]	3.26±0.04 [b]	3.50±0.03 [a]	3.39±0.03 [b]	3.26±0.04 [c]	3.44±0.04 [a]	3.29±0.03 [b]	3.19±0.04 [c]
BW 14 mo	428.57±2.56 [a]	398.41±2.20 [b]	371.77±2.93 [c]	450.47±1.87 [a]	413.68±1.60 [b]	373.63±2.19 [c]	432.61±2.27 [a]	398.21±1.98 [b]	365.38±2.60 [c]
Services per conception (heifers)	1.45±0.38	1.26±0.34	1.40±0.35	1.90±0.15	1.77±0.13	1.91±1.18	2.19±0.18	2.12±0.14	2.28±0.20
AFC, d	707.40±10.32 [b]	713.24±9.65 [b]	724.11±9.61 [a]	715.79±4.04 [b]	721.55±3.47 [b]	753.02±4.73 [a]	712.64±4.22 [b]	718.45±3.67 [b]	734.06±4.82 [a]
First lactation									
Milk, kg (first 100 d)	3,013±126.16 [ab]	3,069±115.28 [a]	2,902±116.50 [b]	2,991±50.67 [a]	2,972±43.73 [a]	2,842±61.93 [b]	3,087±108.10	3,167±98.34	3,157±132.62
Protein, % (first 100 d)	3.07±0.04 [a]	3.02±0.04 [b]	3.01±0.04 [b]	3.08±0.02	3.06±0.02	3.06±0.02	3.11±0.02 [a]	3.08±0.02 [ab]	3.06±0.02 [b]
Fat, % (first 100 d)	3.90±0.11 [a]	3.85±0.10 [b]	3.73±0.1 [b]	3.93±0.04 [a]	3.88±0.04 [b]	3.75±0.05 [a]	3.74±0.05 [a]	3.70±0.04 [a]	3.52±0.06 [b]
Milk, kg	9,041±555.84 [a]	9,117±549.23 [a]	8,596±536.77 [b]	8,984±189.19 [a]	8,901±170.35 [a]	8,252±212.07 [b]	9,275±223.82 [a]	9,289±220.57 [a]	8,811±251.16 [b]
Protein, %	3.21±0.07 [a]	3.13±0.07 [ab]	3.13±0.07 [b]	3.26±0.02	3.23±0.02	3.23±0.03	3.27±0.03 [a]	3.21±0.03 [b]	3.20±0.03 [b]
Fat, %	3.71±0.16 [a]	3.65±0.16 [b]	3.56±0.16 [b]	3.79±0.06 [a]	3.73±0.05 [ab]	3.68±0.07 [b]	3.64±0.07 [a]	3.63±0.06 [b]	3.48±0.07 [b]
Services per conception	2.15±0.48	2.15±0.43	1.98±0.43	2.60±0.22 [a]	2.24±0.17 [ab]	1.83±0.25 [b]	1.84±0.29	1.97±0.26	1.65±0.36
Open days, d	155.92±19.25	139.88±17.43	149.82±17.45	142.88±9.61 [a]	126.32±8.19 [b]	122.76±10.97 [b]	126.75±9.50	111.84±7.48	114.93±10.42
Calving interval, d	433.26±20.87 [a]	413.68±18.79 [b]	426.30±19.10 [a]	421.65±9.45 [a]	401.34±7.73 [b]	399.82±10.56 [b]	403.95±9.19 [a]	383.52±7.02 [b]	390.75±9.54 [ab]
Second lactation									
Milk, kg	10,050±589.48 [a]	9,878±550.76 [ab]	9,421±547.38 [b]	10,401±237.96	10,095±204.53	9,368±273.51	10,880±252.81 [a]	10,744±199.09 [a]	10,223±282.44 [b]
Protein, %	3.22±0.06	3.20±0.06	3.25±0.06	3.23±0.03	3.24±0.02	3.28±0.03	3.25±0.03	3.24±0.02	3.26±0.03
Fat, %	3.70±0.14	3.68±0.13	3.67±0.13	3.62±0.06	3.70±0.05	3.64±0.07	3.58±0.06	3.61±0.05	3.52±0.07
Open days, d	159.59±28.95	146.06±27.18	138.12±26.91	154.44±12.12 [a]	157.59±10.56 [a]	132.45±14.11 [b]	121.77±13.27	132.92±12.07	115.21±16.28
Calving interval, d	414.13±12.29	396.89±8.89	386.81±12.19	420.62±14.93	413.39±14.28	393.93±17.33	391.86±15.09	395.25±14.31	381.77±18.47
Third lactation									
Milk, kg	11,351±929.61 [a]	10,845±885.53 [a]	10,008±847.61 [b]	11,029±340.52 [a]	10,964±312.30 [a]	9,915±416.21 [b]	11,384±354.13 [a]	11,082±277.65 [a]	10,095±385.26 [b]
Protein, %	3.12±0.10	3.11±0.10	3.13±0.09	3.20±0.04	3.19±0.03	3.21±0.05	3.23±0.04	3.23±0.03	3.22±0.04
Fat, %	3.53±0.25	3.52±0.24	3.55±0.23	3.64±0.10	3.67±0.09	3.61±0.12	3.56±0.10	3.54±0.08	3.57±0.11
Milk, kg (LP)	10,243±718 [a]	9,915±684 [a]	9,315±655 [b]	9,957±258 [a]	9,992±236 [a]	9,012±315 [b]	10,366±276 [a]	10,320±216 [a]	9,529±300 [b]
Protein, % (LP)	3.16±0.09	3.12±0.08	3.11±0.08	3.21±0.03	3.18±0.03	3.21±0.04	3.25±0.03	3.19±0.03	3.19±0.04
Fat, % (LP)	3.73±0.22	3.65±0.21	3.65±0.20	3.75±0.08	3.73±0.08	3.63±0.10	3.63±0.08	3.52±0.06	3.50±0.09

Within a row, mean values and standard deviation related to the same explanatory variable with different superscript alphabets are significantly different (P <0.05). ADG in kg/d; LP = lifetime average production per lactation.

Table 6 - Effects of AFC, BCS, and BW on production and reproduction outcomes

Item	AFC (≥751)	AFC (750 – 700)	AFC (≤699)	BCS 14 mo of age (≥3.75)	BCS 14 mo of age (3.5 – 3.25)	BCS 14 mo of age (≤3)	BW 14 mo of age (≥420)	BW 14 mo of age (419 – 380)	BW 14 mo of age (≤379)
n	308	444	364	90	268	172	343	302	135
BCS 14 mo of age	-	-	-	-	-	-	3.48±0.02 [a]	3.33±0.02 [b]	3.19±0.03 [c]
BW 14 mo of age	-	-	-	431.92±3.84 [a]	417.42±2.05 [b]	394.77±2.78 [c]			
Services per conception (heifers)	-	-	-	1.81±0.20	1.87±0.12	2.02±0.15	1.94±0.11	1.91±0.11	1.81±0.16
AFC, d	-	-	-	724.78±7.14	727.58±3.81	735.52±5.17	721.07±3.39 [c]	736.26±3.42 [b]	757.95±5.25 [a]
First lactation									
Milk, kg (first 100 d)	3,046±32.29 [a]	2,961±27.36 [b]	2,917±32.04 [b]	2,886±63.93	2,965±34.52	2,972±46.11	3,001±32.65 [a]	2,967±34.35 [ab]	2,881±51.78 [b]
Protein, % (first 100 d)	3.07±0.01 [a]	3.09±0.01 [a]	3.11±0.01 [a]	3.12±0.02	3.10±0.01	3.08±0.02	3.09±0.01	3.08±0.01	3.08±0.2
Fat, % (first 100 d)	3.93±0.03 [a]	3.82±0.03 [b]	3.83±0.03 [b]	3.98±0.06 [a]	3.84±0.03 [b]	3.74±0.04 [c]	3.92±0.03 [a]	3.85±0.03 [a]	3.71±0.05 [b]
Milk, kg	8,946±133.25	8,816±129.85	8,800±136.95	8,516±234.74	8,848±166.58	8,806±194.68	8,977±139.59 [a]	8,822±141.04 [a]	8,346±180.70 [b]
Protein, %	3.24±0.02	3.24±0.02	3.26±0.02	3.32±0.03 [a]	3.25±0.02 [b]	3.23±0.03 [b]	3.26±0.02	3.23±0.02	3.22±0.02
Fat, %	3.81±0.04	3.75±0.04	3.79±0.04	3.89±0.07 [a]	3.79±0.05 [b]	3.70±0.06 [a]	3.80±0.04 [a]	3.79±0.04 [a]	3.70±0.06 [b]
Services per conception	2.25±0.15	2.24±0.11	2.21±0.14	1.86±0.26	2.19±0.16	2.33±0.23	2.35±0.14	2.23±0.14	1.91±0.23
Open days, d	145.78±7.24 [a]	139.35±6.29 [b]	132.38±6.96 [b]	105.87±13.2 [b]	138.84±7.62 [a]	153.81±10.0 [a]	140.80±6.84	132.51±7.11	136.43±10.28
Calving interval, d	418.82±7.78	408.48±6.10	407.80±6.99	388.27±13.10 [b]	414.11±7.39 [a]	428.23±10.58 [a]	415.77±7.04	403.92±6.91	411.44±10.01
Second lactation									
Milk, kg	10,221±196.81	10,220±153.83	10,449±173.75	9,648±311.66 [b]	10,314±193.37 [a]	10,086 ±260.53 [ab]	10,519 ±177.44 [a]	10,219±190.68 [a]	9,671±253.10 [b]
Protein, %	3.28±0.02	3.24±0.02	3.24±0.02	3.34±0.04 [a]	3.26±0.02 [b]	3.27±0.03 [ab]	3.24±0.02	3.25±0.02	3.24±0.03
Fat, %	3.71±0.05	3.67±0.04	3.65±0.04	3.68±0.08	3.62±0.05	3.56±0.07	3.61±0.04	3.69±0.05	3.61±0.06
Open days, d	159.17±10.21	147.54±8.63	150.12±9.07	135.91±16.62	147.61±10.73	141.88±14.73	157.40±9.46 [a]	156.51±9.64 [a]	126.08±12.4 [b]
Calving interval, d	431.87±10.87 [a]	409.58±8.93 [b]	430.73±10.29 [a]	432.94±16.40	423.17±10.48	407.50±15.07	433.43±9.58 [a]	417.64±9.53 [ab]	394.75±13.57 [b]
Third lactation									
Milk, kg	9,903±281.97 [b]	10,570±237.23 [a]	10,922±266.26 [a]	10,710±456.97	10,577±312.19	10,695±428.53	11,057±246.53 [a]	10,443±262.25 [b]	9,253±345.12 [c]
Protein, %	3.22±0.02 [a]	3.15±0.02 [b]	3.17±0.03 [ab]	3.25±0.05	3.24±0.04	3.22±0.05	3.18±0.03	3.18±0.03	3.18±0.04
Fat, %	3.75±0.07 [a]	3.61±0.06 [b]	3.61±0.07 [b]	3.63±0.12	3.59±0.08	3.55±0.11	3.65±0.07	3.64±0.07	3.65±0.09
Milk, kg (LP)	9,602±213	9,861±179	9,859±201	9,454±347	9,797±237	9,952±326	10,109±181 [a]	9,758±193 [a]	8,760±254 [b]
Protein, % (LP)	3.23±0.03 [a]	3.17±0.02 [a]	3.18±0.02 [b]	3.29±0.05	3.21±0.03	3.20±0.04	3.20±0.02	3.19±0.03	3.18±0.03
Fat, % (LP)	3.79±0.07 [a]	3.67±0.06 [b]	3.71±0.06 [ab]	3.75±0.11	3.69±0.08	3.55±0.10	3.73±0.06	3.73±0.06	3.65±0.08

Within a row, mean values and standard deviation related to the same explanatory variable with different superscript alphabets are significantly different (P <0.05). AFC = average age at first calving; BCS = body condition score (5 point scale); BW = body weight (kg); LP = lifetime average production per lactation.

5.1.1 Average daily weight gain during the heifer rearing period and production performance

A significant difference (P <0.05) between the highest and the lowest ADG evaluated periods was found in relation to the AFC. The shortest AFC was achieved with the highest overall ADG ($\geq$ 0.950 kg/d) and prepubertal growth ($\geq$ 0.970). The AFC declined by 17, 38, and 22 d when highest ADG in prepubertal, postpubertal, and total growth, respectively. Milk yield in the first, second and third lactation and in LP of all ADG groups $\leq$ 0.849 kg/d was significantly lower (P <0.05) than in other groups (Table 5). The highest milk yield in the first lactation was found in medium group of ADG (0.969 to 0.850 kg/d of prepubertal growth; 0.949 to 0.850 of total growth) except for postpubertal growth. The milk yield reached 9,117 kg/305 d per cow in prepubertal growth in accordance with the total growth intensity measured from 5 to 14 mo of age 9,289 kg/305 d per cow (Table 5). The same trend was observed in milk yield in the first 100 d in the first lactation. The regression coefficients and correlation for milk yield in the first lactation and total ADG were 1,835 kg (P <0.05) and 0.09 (P <0.05), respectively. The correlation between evaluated variables was low. The highest milk yield in the LP (Table 5) occurred in the group with the highest ADG ($\geq$0.950 kg/d; in prepubertal growth $\geq$0.970) except for postpubertal growth. The differences in milk yield ranged from 837 to 945 kg/305 d per cow (P <0.05) compared to the lowest ADG ($\leq$0.849 kg/d) during the entire period observed. The regression and correlation coefficients in the milk yield of LP and total ADG were 3,865 kg (P <0.01) and 0.27 (P <0.01), respectively. The level of ADG affected the content of milk components, the higher the ADG, the higher the percentage of fat and protein content. Similar differences (P <0.05) in milk yield were determined even in the second and third parity (Table 5). Groups with the highest ADG showed the highest milk yield in the second and third parity except for the postpubertal growth group in the third parity. Open days in the first and the second parity differed (P <0.05) only in relation to the intensity of postpubertal growth. Longer open days were observed with higher ADG. The differences between high ADG ($\geq$0.950 kg/d) and low ADG ($\leq$0.849 kg/d) in open days were 20 d for the first parity and 22 d for the second parity. Calving interval between the first and second calving reached 414 d in the medium ADG (0.969 to 0.850 kg/d) compared to 433 d in the high ADG ($\geq$ 0.970 kg/d) and 426 d in the low ADG ($\leq$ 0.849 kg/d) (P <0.05). Similar results of ADG impacts but with lower differences were observed between the other evaluated periods (Table 5).

The Figure 6 shows the evaluation of probability for cull in the first parity depending on overall ADG during the rearing period. The higher the ADG, the higher the probability for cull. Based on the Figure 6 bellow, ADG 0.800 kg/day can be regarded as risky, for the probability for cull is over 80%.

Figure 6 – Relationship between probability for cull in the first parity and average daily weight gain (kg/day) - ADG

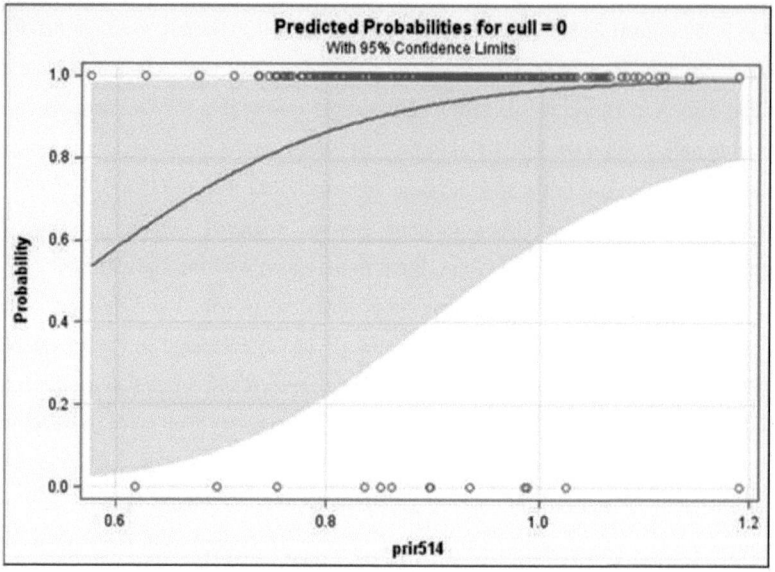

Prir514 = overall average daily gain per rearing period (5 – 14 months of age)

5.1.2 Body weight and body condition at 14 month of age and production performance

The higher the BW at 14 mo of age ($\geq$420 kg), the higher the milk yield in the first (8,977 kg), second (10,519 kg), third (11,057 kg) parity, and average LP (10,109 kg) (P <0.05). The higher the BW at 14 mo of age, the lower the AFC (721 d) (P <0.05) (Table 6). The evaluated groups of BCS at 14 mo showed no differences between AFC or milk yield in the first lactation and average LP. However, the components in milk (fat and protein) in the first parity differed (P <0.05) according to the level of BCS at 14 mo of age. The higher the BCS ($\geq$ 3.75point), the higher the content of milk components detected (protein 3.32 % and fat 3.89 %). A significant difference (P <0.05) in the open days in the first parity (-48 d) and calving interval between first and second parity (-40 d) was found between the highest BCS ($\geq$ 3.75 point) and the lowest BCS ($\leq$3 point).

The regression and correlation coefficients for the relationships between BCS and prepubertal ADG (5 to 10 mo of age) were 1.13 points of BCS (P <0.001) and 0.33 (P <0.001), respectively; BCS and postpubertal ADG (11 to 14 mo of age) were 1.91 points of BCS (P <0.001) and 0.45 (P <0.001), respectively; and BCS and total ADG (5 to 14 mo of age) were 1.64 points of BCS (P <0.001) and 0.41 (P <0.001), respectively. Postpubertal growth calculated from 11 and 14 mo of age had the greatest influence on BCS and showed a medium level of correlation.

5.1.3 Age at first calving

The distribution of AFC is shown in Figure 4. The overall mean and SD of AFC was 729±58.54 d (Table 1). The AFC level had a significant impact on milk yield in the first 100 d of first lactation (P <0.05). The group of cows with AFC $\geq$751 d produced 3,046 kg of milk/100 d, whereas the group with AFC $\leq$699 d produced only 2,917 kg/100 d (Table 6). The highest milk fat content, which was found at the highest AFC ($\geq$751 d), significantly differed from the other groups (P <0.05). Milk protein was significantly the lowest at this level of AFC. There were no differences in milk yield and milk components content for the second parity. The groups differed in the third parity, showing the highest milk yield (10,922 kg; P<0.05) for the lowest AFC group accompanied by a lower level of milk protein and fat (3.17%, 3.61%, respectively; P>0.05). The lowest milk yield in LP occurred with an AFC $\geq$751 d. Open days in the first parity were 14 d longer in group with AFC $\geq$751 d than in group with AFC $\leq$699 d and 6 d longer than in group with AFC 750-700 d (P <0.05).

Growth from 5 to 18 mo of age according to AFC is shown in Figure 7. The ADG increased until 10 mo of age when it reached its maximum (AFC ≥751 d: 0.96±0.08 kg/d; AFC from 750 to 700 d: 0.90±0.10 kg/d, and AFC ≤699 d: 0.89±0.12 kg/d) and then followed a gradual decline.

Figure 7 - Growth of heifers (average daily weight gain in kg/d - ADG) according to age at first calving (AFC).

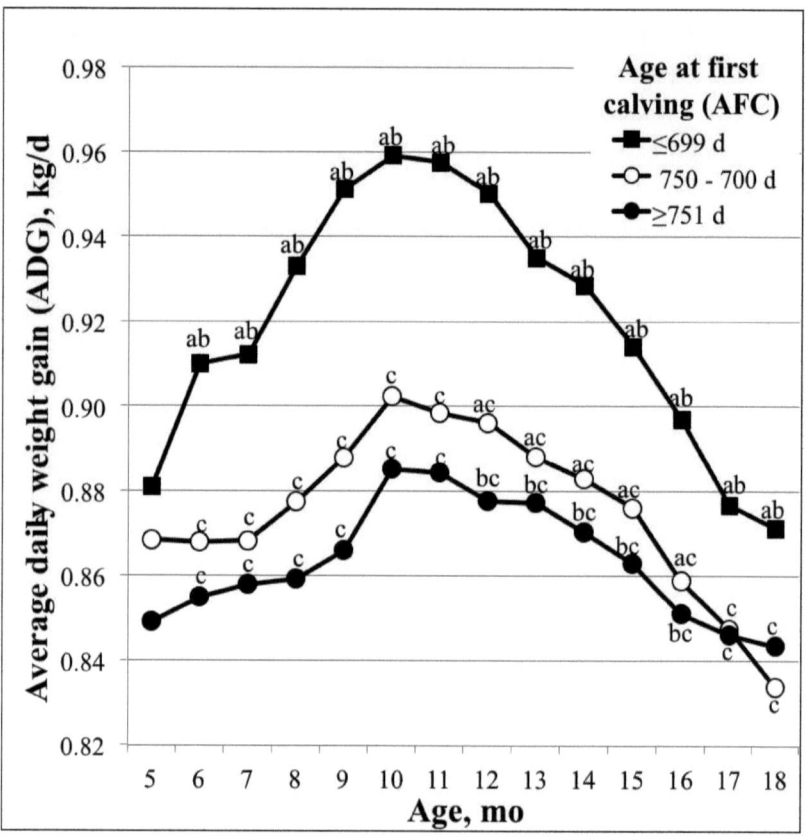

Ages at calving were grouped as: high (a) = ≥ 751 d, n = 308, mean± SD = 804,18±48.78 d; medium (b) = 750 to 700 d, n = 444, mean = 723.07±14.63 d; and low (c) =≤699 d, n = 364, mean = 673.89±19.22 d. Differences between groups (P <0.001) are denoted with different letters vertically.

Changes in BCS during the postpubertal period of growth according to AFC are described in Figure 8. The BCS ranged between 3.2 and 3.5 points from 11 to 16 mo of age. The mean and SD at 14 mo of age (approximate conception period) were 3.36±0.35 points for AFC ≥751 d; 3.33±0.32 points for AFC 750-700 d; and 3.42±0.34 points for AFC ≤699 d. The BCS reached 3.5 points for the group AFC ≤699 d and ≥751 d at 17 mo of age.

Figure 8 - Body condition score (BCS) in the postpubertal period of growth according to age at first calving (AFC).

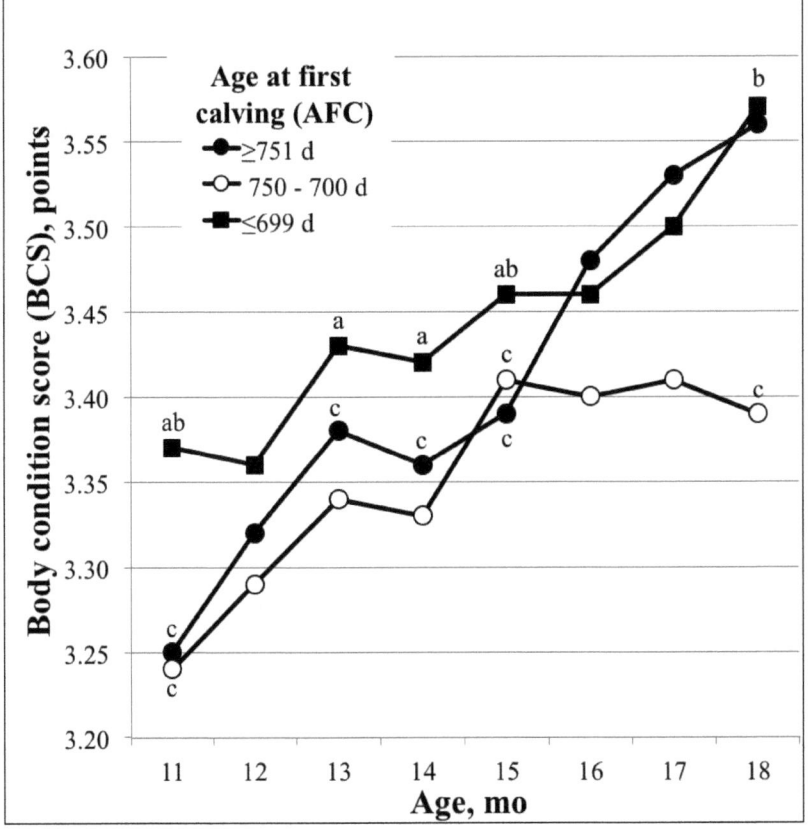

Ages at calving were grouped as: high (a) = ≥ 751 d, n = 308, mean = 804,18±48.78 d; medium (b) = 750 to 700 d, n = 444, mean = 723.07±14.63 d; and low (c) =≤699 d, n = 364, mean = 673.89±19.22 d. Differences between groups (P <0.05) are denoted with different letters vertically.

The regression and correlation coefficients for the relationship between AFC and prepubertal ADG (5 to 10 mo of age) were -59.94 d (P <0.01) and -0.14 (P <0.01), respectively; AFC and postpubertal ADG (11 to 14 mo of age) were -180.03 d (P <0.001) and -0.32 (P <0.001), respectively; and AFC and total ADG (5 to 14 mo of age) were -99.90 d (P <0.001) and -0.20 (P <0.001), respectively. The postpubertal growth had the highest impact on AFC.

5.2 Evaluation of heifer growth from the perspective of dairy production economics

The objective of the next part of this dissertation was to analyse the effect of age at first calving (**AFC**), average daily weight gain (**ADG**) and herd milk yield (**MY**) level on selected production, reproduction and economic parameters. Management decisions on herd replacement can have a significant impact on profitability of a farm as a whole. A lower AFC does not always lead to higher profitability, depending on a dairy farm conditions and management system. Optimum herd replacement rates do not assure the highest profitability because they are considerably influenced by herd milk yield and market prices of inputs and outputs. Chapter 4.1.2 contains the data and calculation procedures. Tables 7 and 8 and Figures 9, 10, 11 and 12 in the following chapter list the results obtained.

Table 7 - Effect of age at first calving (AFC, d), average daily weight gain (ADG, kg/d) and level of milk yield (MY, kg/yr) on life production parameters (33 farms, 2011)

Item		AFC (≥800)	AFC (799–750)	AFC (≤749)	ADG (≥0.800)	ADG (0.799–0.700)	ADG (≤0.699)	MY(≥8500)	MY (8499–7000)	MY(≤6999)	Mean	SD
n (farms)		13	12	8	7	12	14	12	11	8	33	33
MY, kg		7,327±337	8,129±377	8,126±459	8,132±409[a]	8,060±327[a]	7,110±324[b]	-	-	-	7,989.03	1,324.99
Protein, %		3.37±0.04[b]	3.43±0.04[b]	3.54±0.05[a]	3.45±0.04	3.41±0.04	3.43±0.04	3.41±0.04	3.42±0.04	3.41±0.04	3.40	0.13
Fat, %		3.95±0.06[a]	3.86±0.07[ab]	3.71±0.08[b]	3.72±0.07[b]	3.93±0.06[a]	3.95±0.06[a]	3.90±0.06	3.95±0.06	3.93±0.07	3.87	0.21
Conception rate after insemination (heifers)	1st	-	-	-	56.02±2.79[b]	58.02±2.27[b]	65.22±2.21[a]	-	-	-	60.51	7.94
Conception rate after inseminations (heifers)	all	-	-	-	55.05±3.91[b]	57.70±3.18[b]	64.07±3.10[a]	-	-	-	60.21	11.27
Conception rate after insemination (cows)	1st	42.95±2.19[a]	40.16±2.56[ab]	35.70±3.00[b]	39.74±2.86	38.52±2.36	43.56±2.30	38.20±2.99	40.23±2.70	43.57±3.35	39.87	8.37
Conception rate after inseminations (cows)	all	48.88±2.70[a]	39.80±3.01[b]	32.78±3.69[b]	41.51±3.89	42.41±3.18	44.44±3.10	38.54±3.74	44.78±3.39	46.20±4.22[b]	42.10	11.72
Culling of cows, %		31.01±3.20[b]	33.42±3.60[b]	40.59±4.36[a]	35.91±4.39	31.11±3.52	33.06±3.50	37.70±4.07[a]	35.53±3.84[a]	26.47±4.69[b]	34.78	10.43
Services per conception		2.02±0.09[b]	2.23±0.10[ab]	2.36±0.12[a]	2.13±0.11[ab]	2.27±0.09[a]	2.06±0.09[b]	2.21±0.11[ab]	2.28±0.10[a]	1.98±0.12[b]	2.22	0.35
Calving interval, d		401.23±5.08[ab]	396.35±5.64[b]	415.65±6.97[a]	401.09±7.19	399.22±5.74	404.74±5.74	386.43±6.54[b]	402.90±6.03[a]	412.24±7.17[a]	405.77	17.63
Days open, d		119.84±5.18[ab]	105.41±6.01[b]	131.51±7.02[a]	115.15±8.12	118.64±6.63	119.00±6.32	106.11±7.60[b]	116.15±7.03[ab]	128.82±8.72[a]	120.74	20.48
ADG, kg/d		0.70±0.05[b]	0.76±0.05[b]	0.89±0.07[a]	-	-	-	0.79±0.03[a]	0.75±0.03[a]	0.64±0.03[b]	0.73	0.15
AFC, d		-	-	-	782.48±11.51	786.13±9.24	793.27±9.17	780.03±10.69[b]	792.59±10.06[a]	794.22±12.28[a]	777.95	37.50
Number of lactations		2.67±0.10[a]	2.42±0.11[ab]	2.29±0.13[b]	2.24±0.12[b]	2.68±0.09[a]	2.56±0.09[a]	2.45±0.13	2.48±0.12	2.67±0.15	2.40	0.43
Death rate of calves, %		5.51±1.04	4.18±1.18	5.94±1.40	4.63±1.35	4.49±1.06	6.35±1.04	2.43±0.95[b]	2.65±0.90[b]	10.34±1.12[a]	5.08	3.71
Total loss of calves, %		10.81±1.34	11.45±1.52	12.73±1.81	9.73±1.71	11.21±1.34	12.86±1.32	8.33±1.35[b]	8.10±1.28[b]	17.15±1.60[a]	11.62	4.74

Within a row, mean values and standard deviation related to the same independent variable with different superscript alphabets are significantly different (P < 0.05).

42

Table 8 - Effect of age at first calving (AFC), average daily weight gain (ADG), and level of milk yield (MY, kg/yr) on rearing costs and profitability (33 farms, 2011)

Item	AFC (≥800)	AFC (799 - 750)	AFC (≤749)	ADG (≥0.800)	ADG (0.799 - 0.700)	ADG (≤0.699)	MY (≥8500)	MY (8499 - 7000)	MY (≤6999)	Mean	SD
N (farms)	13	12	8	7	12	14	12	11	8	33	33
TP, CZK/l	-0.83±0.55	-0.35±0.63	-1.03±0.74	-0.02±0.70	-0.69±0.55	-1.30±0.54	0.15±0.67[a]	-0.37±0.63[ab]	-1.77±0.79[b]	-0.62	1.76
TP, CZK/per cow	-5.428±3.982	-2.868±4.472	-7.151±5.425	-800±5.062	-4.561±3.977	-8.186±3.923	664±4.851[a]	-2.927±4.568[ab]	-12.495±5.594[b]	-4.074.47	12.440.84
TP, CZK/feeding d	-14.87±10.82	-7.80±12.17	-19.50±14.73	-2.32±13.76	-12.32±10.80	-22.55±10.65	1.87±13.18[a]	-8.22±12.41[ab]	-34.11±15.21[b]	-11.20	33.88
PROF, %	-5.71±8.52	2.42±9.69	-9.66±11.52	6.21±11.16	-5.23±8.71	-10.31±8.59	2.67±10.81[a]	0.91±10.26[ab]	13.91±12.81[b]	-2.38	27.17
Cow depreciation costs, CZK/cow	5.829±650[b]	6.271±739[b]	8.275±879[a]	7.675±817[b]	7.119±656[b]	4.832±651[a]	7.458±830	6.397±787	5.139±984	6.383	2.491
Cow depreciation costs, CZK/feeding d	15.93±1.77[b]	17.21±2.01[b]	22.68 ±2.39[a]	21.13±2.22[b]	19.45±1.78[b]	13.19±1.77[a]	20.42±2.26[a]	17.53±2.15	14.06±2.68	30.47	79.66
Cow depreciation costs, CZK/L	0.91±0.09[b]	0.84±0.11[b]	1.09±0.13[a]	1.01±0.12[b]	1.01±0.10[b]	0.76±0.10[a]	0.95±0.12	0.91±0.12	0.87±0.15	1.54	4.09
Milk and milk replacer costs, CZK/heifer	1.759±431[ab]	1.249±504[b]	2.723±645[a]	3.051±743[a]	1.670±452[ab]	1.156±441[b]	-	-	-	1.785	1.416
Total feed costs, CZK/heifer	10.079±1.291	10.758±1.511	9.790±1.931	12.228±2.194	9.894±1.333	9.503±1.300	-	-	-	10.263	3.902
Total costs, CZK/heifer (6 to 21 mo of age)	19.980±3.756	26.009±4.272	25.389±5.081	20.183±4.883	26.670±3.833	21.067±3.780	-	-	-	22.660	11.914
Total costs, CZK/per calf (6 mo age)	14.905±2.036	11.633±2.098	10.233±2.816	11.244±3.043	13.377±2.183	13.384±2.324	-	-	-	13.145	5.947

Within a row, mean values and standard deviation related to the same independent variable with different superscript alphabets are significantly different (P < 0.05).
TP = total profit without subsidies; PROF = profitability without subsidies; 1USD = 20CZK.

5.2.1 Growth intensity and AFC

The regression coefficients and correlation for AFC and ADG were -35.78 d ($P < 0.24$) and 0.34 ($P < 0.05$), respectively. AFC and ADG were analysed separately (Tables 1 and 2), although only modest correlation was found between them. Average daily gain differed ($P < 0.05$) between the groups of the highest AFC (≥ 800 d) and lowest AFC (≤ 749 d) with means of 0.70 kg and 0.89 kg, respectively. Dividing the ADG into groups evidenced that the highest ADG was associated with the lowest AFC, but the difference between the high (≥ 0.800 kg) and low (≤ 0.699 kg) ADG group means was only between 10 and 11 d and was not significant (Table 7).

All of the evaluated reproduction traits differed significantly ($P < 0.05$) among AFC groups (Table 7). The conception rate after first and all inseminations for cows was highest in herds of the group AFC ≥ 800 d. The highest AFC (≥ 800 d) and lowest AFC (≤ 749 d) groups had conception rates of 43% and 36% after first insemination and 49% and 33% after all inseminations, respectively. A similar result was seen in the number of services per conception, with only the groups AFC ≥ 800 d and AFC ≤ 749 d differing significantly from each other ($P < 0.05$; Table 7). The shortest mean days open (105 d) and calving interval 396 d were found in the middle AFC group (799 to 750 d) and the longest days open (132 d) and calving interval (416 d) in the lowest AFC group (≤ 749 d) ($P < 0.05$). The differences between the AFC groups were equivalent to approximately one oestrous cycle for heifers (Table 7). The regression coefficients and correlations for AFC and calving interval were -0.16 d ($P < 0.03$) and -0.36 ($P < 0.04$) and for AFC and days open -0.19 d ($P < 0.01$) and -0.39 ($P < 0.03$), respectively. Differences in MY between AFC groups were not significant and similar MY mean values were for the lowest (8,126 kg) and middle (8,129 kg) AFC groups. Mean lactations completed differed significantly ($P < 0.05$) among the AFC groups, with the highest mean being for the highest AFC group, at 2.67 lactations, and the lowest value being for the lowest AFC group, at 2.29 lactations. Similar results were seen for the ADG groups. The group of highest ADG (which had the correspondingly lowest AFC) achieved 2.24 completed lactations. The analysis shows that those herds that reared more intensively (having low AFC and high ADG) had lower conception rates among heifers after first and all inseminations. The differences between the group of highest ADG (≥ 0.800 kg) and lowest ADG (≤ 0.699 kg) were around 10 percentage points for both aforementioned evaluated variables. The differences between culling of heifers and cows according to AFC are shown in Table 7 and Figure 9. The highest culling rate for cows (41%) was found in the group with lowest AFC ($\leq$749 d) (P < 0.05). Figure 9 shows that the main reasons for culling of heifers were fertility problems and the highest percentage was observed in the group with the highest AFC ($\geq$800 d). Reproductive problems tended to be the main reasons also for culling of cows, and other major reasons included movement disorders, mammary gland diseases, low production, and, in the group with highest AFC ($\geq$800 d), digestive diseases.

Figure 9. Reasons for culling[1] of heifers (left panels) and cows (right panels) according to age at first calving (AFC)[2] (upper panels) and milk yield (MY)[3] (lower panels) (33 farms, 2011)

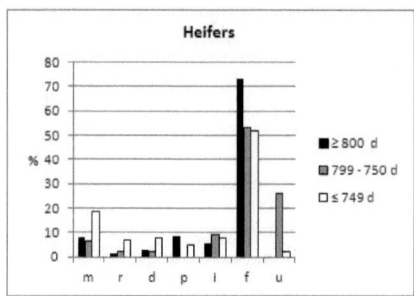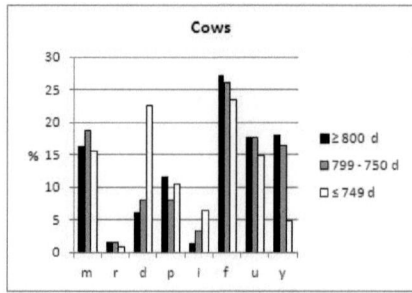
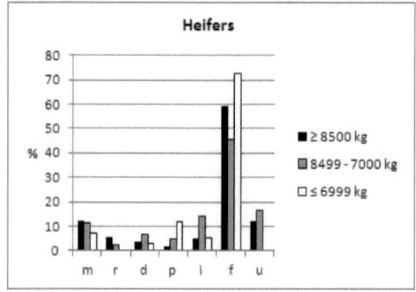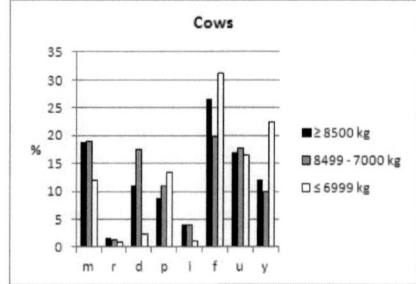

[1]m = movement disorders, r = respiratory diseases, d = digestive diseases, p = postpartum complications, i = injuries, f = low fertility, u = mammary gland diseases, y = low milk production.
[2]AFC was grouped as high: $\geq$800 d, n = 13 farms, mean = 816.96 $\pm$ 20.69 d, median = 810 d; middle: 799 to 750 d, n = 12 farms, mean = 765.00 $\pm$ 13.32 d, median = 761 d; and low: $\leq$749 d, n = 8 farms, mean = 733.97 $\pm$ 9.47 d, median = 735 d.
[3]MY was grouped as high: $\geq$ 8,500 kg, n = 12 farms, mean = 9,197 $\pm$ 748 kg, median = 8,838 kg; middle: 8,499 to 7,000 kg, n = 11 farms, mean = 7,679 $\pm$ 436 kg, median = 7,555 kg; and low: $\leq$6,999 kg, n = 8 farms, mean = 6,286 $\pm$ 586 kg, median = 6,335 kg.

Figure 10 shows subjective evaluations by farmers of effects of some factors on heifer rearing success. Nutrition and feeding were rated the highest (4.5 points). Obviously, the feeding philosophy in heifers is different from the one in cows. Farmers fed diets with high nutrient concentrations to lactating cows to avoid limited production. On the contrary, heifers should receive diets with precisely defined nutrient contents for different rearing phases. The health is regarded by farmers as equally important (4.4 points). Cowman care of heifers is in the third place (3.8 points). Other factors which the farmers considered to have a significant impact on successful rearing included hygiene and EU subsidy policies.

Figure 10 – Effect of selected factors on heifer rearing success (subjective evaluation)[1] (33 farms, 2011)

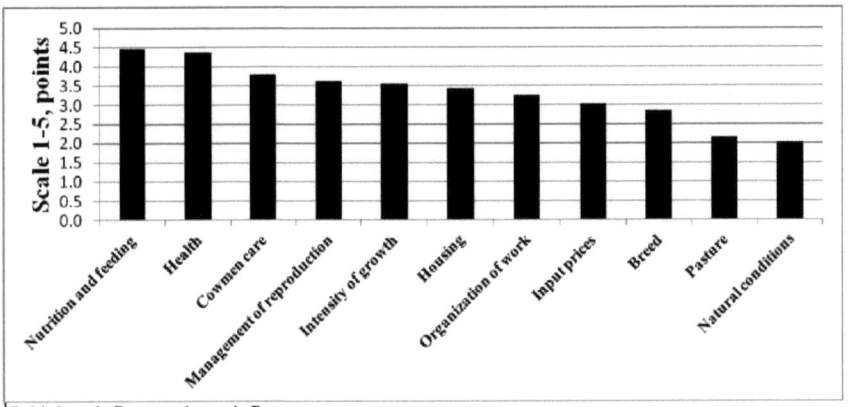

[1]5=highest influence, 1= no influence.

5.2.2 Heifer rearing and farm profitability

Economic consequences of heifer rearing strategies with different intensities (parameters ADG and AFC) are summarized in Table 8. Statistically significant differences (P <0.05) were found only in feed costs (milk and milk replacers) and in cow depreciation. The highest cost of milk and milk replacers was found in the group with the lowest AFC ≤ 749 days, amounting to 2,723 CZK per heifer. Similar results were observed in the group with the highest ADG ≥ 0.800 kg: the cost of milk and milk replacers was 3, 051 CZK per heifer, which resulted in lower AFC. Heifer feed costs and total costs did not show any significant differences between the groups and mention should also be made of a great variability among the farms. Figure 11 shows percentages of different feed costs of total costs.

Figure 11 – Feed cost proportions of total costs per raised heifer (33 farms, 2011)

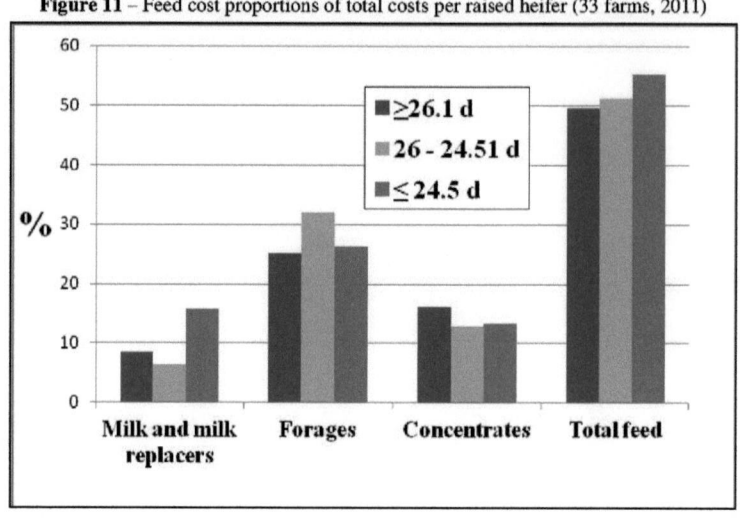

Total cost of a raised heifer ranged between 20,000 CZK and 27,000 CZK in all the groups under evaluation (Table 8). Based on Table 3 in Chapter 4.1.2., we can estimate the losses incurred by prolongation of the rearing period by one day on feed as 22 CZK, but with a considerable variation. It means that each time the rearing period is prolonged by one cycle (21 days) over the determined optimum term, total costs are increased approximately by 462 CZK.

The highest cow depreciation cost was found in the group with the lowest AFC ≤ 749 days, amounting to 8,275 CZK per cow. This group also had the highest culling rate of 41% (Table 8). The highest cow depreciation was also found in the cow group with the highest ADG ≥ 0.800 kg, amounting to 7,675 CZK per cow (P<0.05). Regression and correlation coefficients between ADG and cow depreciation were -51.97 CZK per cow (P <0.001) and

0.51 (P <0.01), respectively. The heifer rearing intensity characterized by medium AFC (799 - 750 days) and ADG (0,799 – 0,700 kg) groups was determined as the most profitable approach to heifer rearing. These groups showed the highest values of parameters characterizing total farm profit excluding subsidies and level of profitability excluding subsidies of 2.42%, despite the fact that these groups had the highest costs during the rearing period (Table 8).

5.2.3 Effect of herd milk production on farm profitability

Analysis of milk production showed that higher MY values lead to poorer conception rates after the first insemination and all inseminations in cows, but the differences were non-significant (Table 7). Group of farms with the highest milk production ≥ 8 500 kg had the lowest conception rates after the first insemination and all inseminations in cows of 38% and 39%, respectively. However, this group of herds had the shortest open days of 106 days and calving interval of 386 days (P <0.05) as compared with herds showing medium milk production level MY 8,499 – 7,000 kg and lowest milk production level MY ≤ 6, 999 kg, but also the highest culling rate (Table 7). Farms with the lowest AFC (780 days) had the highest MY. The difference in AFC between the groups of farms with the highest and lowest MYU was 14 days (P < 0.05). The group of farms with the highest average MY achieved the highest ADG and AFC (P < 0.05). The Figure 12 shows all the 33 farms and the trend curve of AFC with the farms ranked from the highest to lowest average MY.

Figure 12 – Milk yield (thous. kg) and age at first calving (AFC)[1), 2)] (33 farms, 2011)

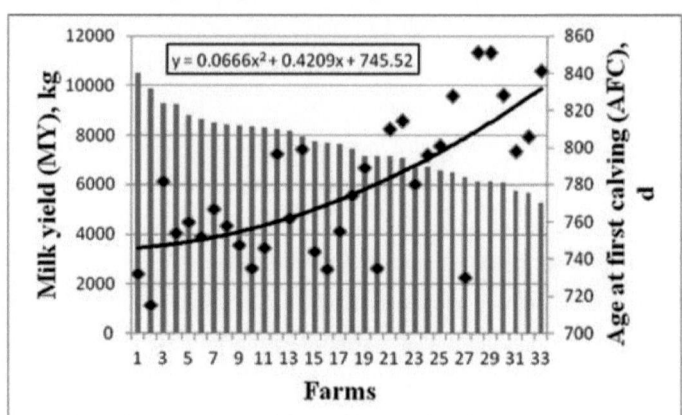

[1] total weight gain 410 kg; [2)] total weight gain from birth to 6 mo of age 193 kg

The analysis of MY showed that higher production led to lower conception rates after first and all inseminations in cows, but the differences between groups were not statistically significant (Table 7). In the group with highest MY (≥8,500 kg), the conception rates of cows after first and all inseminations were the lowest at 38% and 39%, respectively. Nevertheless, this high MY group also had the shortest days open (106 d) and calving interval (386 d) (P <0.05) in comparison with the middle (8,499 to 7,000 kg) and lowest (≤6,999 kg) MY group (Table 7), which had higher culling rates. The highest MY occurred for the lowest AFC group (780 d). The difference in AFC between the high and the low MY groups was 14 d (P < 0.05). The highest MY group had the highest ADG and the lowest AFC (P < 0.05). The herds with higher production (MY ≥ 8,500 kg and MY 8,499 to 7,000 kg) had lower losses of calves (P < 0.05). The differences between the highest and lowest MY groups for death rate and total loss of calves were approximately 8 to 9 percentage points. Differences according to MY in culling of cows are shown in Table 8 and Figure 9. The lowest culling of cows (27%) was found in the lowest MY group. The difference in culling rates between the low and the high MY groups was 10 percentage points and between the low and middle MY groups was 9 percentage points (P < 0.05). Figure 9 shows that the main reason for culling of cows regardless of MY group was fertility problems. Other important reasons included movement disorders, mammary gland diseases, and, for the group with lowest MY, low production.

The highest total profit and level of profitability without subsidies were achieved in the group with the highest MY, and this was statistically significant (P < 0.05) in comparison with the lowest MY group (Table 8). The highest reported cow depreciation costs were found in the highest MY group (Table 8), which was associated with the highest culling rate for cows (38%). The regression coefficient and correlation for MY and total profit were CZK 2.86/cow (P < 0.08) and 0.31 (P < 0.08), respectively.

5.3 Comparison of selected management and economic parameters within the breeds

Within a brief breed evaluation the analysed data were also related to milk performance only. Holstein heifers had the lowest total rearing cost of 21,975 CZK, Czech Fleckvieh had by 2,863 higher heifer rearing cost and mixed breed herds were by 4,978 CZK/head higher (Table 9).

Table 9 – Average parameters of selected cost items in the heifer rearing [1], [2] including profit calculated per litre of sold milk (33 farms, 2011)

Parameter	Breed		
	H	C	CH
n (farms)	17	8	8
Cost of milk and milk replacers, CZK/head	1143	980	969
Cost of forage, CZK/head	6 485	7 236	7 529
Cost of concentrate, CZK/heads	3 125	3 555	4 148
Total feed costs, CZK/head	10 254	11 271	12 146
Total costs, CZK/head	21 975	24 778	26 953
Cow depreciation costs, CZK/L of milk	0.87	1.04	0.72
Profit without subsidies, CZK/L of milk	-0.20	-1.15	-1.00
Profitability without subsidies /with subsidies, %	0.04/ 0.13	-0.07/ 0.04	-0.11/ -0.02

[1]total average daily weight gain 410 kg; [2] total average daily weight gain 193 kg at 6 months

Holstein herds showed generally poorer reproduction performance, the highest average milk production of 8,581 kg milk, the lowest AFC 24.8 months, the highest cow culling rate of about 36 % and the lowest longevity (2.2 lactations). In spite of that they achieved the highest average profit per 1 litre of milk (0.20 CZK) and average profitability without subsidies of 0,04 and average profitability with subsidies of 0.13 CZK (Table 10).

Table 10 – Average values of selected management parameters in heifer rearing and dairy cow production (33 farms, 2011)

Parameter	Breed		
	H	FV	FV and H
ADG, kg/head/day	0.70	0.76	0.77
Conception rate after 1st service, % (heifers)	60.73	63.35	57.25
Conception rate after all services, % (heifers)	58.99	63.63	59.08
AFC, days	757	819	781
MY, kg	8 581	6 643	8 077
Annual milk sales, CZK/cow	8 186	6 229	7 609
Annual milk sales, CZK/DOF (day on feed)	183.57	142.41	176.75
Milk price, CZK/L	8.17	8.38	8.51
Conception rate after 1st insemination, % (cows)	36.12	46.50	40.26
Conception rate after all inseminations , % (cows)	36.05	50.98	44.58
Cow culling rate, %	36.16	33.62	33.02
Services per conception, number	2.40	1.89	2.15
Open days	128	108	108
Calving interval, days	411	397	403
Number of concluded lactations	2.20	2.79	2.44
Number of calves reared per 100 cows, heads	102	98	98
Calf death rate, %	4.47	8.08	3.36
Total calf losses, %	11.64	14.74	8.47

MY=milk production; ADG=average daily weight gain; AFC= first calving age

6 Discussion

6.1 Intensity of rearing dairy heifers and subsequent production

Growth is one of the fundamental processes that takes place during the life of an animal. It is influenced by genetic potential, nutrition and environmental conditions (Mourits et al., 2000). Froidmont et al. (2013) reported that the onset of heifers' sexual maturity depends more on body weight than age and can be significantly influenced by the level of nutrition. According to Bar-Peled et al. (1997) the body's cells and parenchyma cells of the mammary gland develop at similar rates in the period from birth to 3 mo of age, but the parenchyma of the mammary gland develops 3 to 4 times faster than the cells of the body in the period from 3 to 9 mo of age (prepubertal period). Most studies have documented the negative impact of high ADG during the heifers' rearing period on milk production in the first lactation and especially in the prepubertal period due to reduction of parenchyma growth in the mammary gland (Daniels, 2010; Ettema and Santos, 2004; Silva et al., 2002 and Sejrsen and Purup, 1997), although some experiments did not confirm this relationship (Pirlo et al., 2000; Waldo et al., 1998; Kertz et al., 1984), or only in cases where the ADG exceeded 0.7 kg/d (Abeni et al., 2000), 0.9 kg/d (Knight and Sorensen, 2001) or 1.0 kg/d (Mourits et al., 1999a). Van Amburgh et al. (1998) gave an upper limit threshold of 0.7 to 0.8 kg/d for the optimal development of the mammary gland, and it was validated by a meta-analysis of eight studies (Zanton and Heinrichs, 2005). Van Amburgh et al. (1998) reported a decrease in milk production during subsequent lactations from 10 to 40% in relation to low ($\leq$ 0.4 kg/d) or high (> 0.8 kg/d) ADG before puberty. Shamay et al. (2005) concluded that an ADG of about 0.7 kg/d is optimal for achieving maximum performance. In this study, ADG $\geq$ 0.970 kg/d in the prepubertal period (5 to 10 mo of age) and ADG $\geq$ 0.950 kg/d in the total period (5 to 14 mo of age) showed a negative influence on subsequent milk production in the first lactation. The lowest milk yield (P <0.05) in all the evaluated groups was detected in the first lactation with ADG $\leq$ 0.849 kg/d. According to Tozer and Heinrichs (2001) and Sakaguchi et al. (2005), the negative effect of high levels of nutrition before puberty occurs in all breeds. According to Madgwick et al. (2005), accelerated growth of Holstein heifers during puberty supported by a higher concentration of protein in the diet does not reduce subsequent milk production. The same conclusion is supported by Macdonald et al. (2005), who stated that heifers with faster growth during the prepubertal period reduced the development of the mammary glands, but milk yield did not decline because of better physical readiness for the first lactation caused by

52

previous faster growth. Le Cozler et al. (2008) added that heifers with a high potential for milk production appear to be less sensitive to significant negative effects of the high level of nutrition and also that faster growth has a positive effect on the attainment of puberty. In this study, faster growth led to earlier AFC, which was related to earlier maturity of these animals. Foldager and Sejrsen (1991) stated that an increase in ADG from 0.4 to 0.6 kg/d during heifers' rearing led to increased subsequent milk production and mammary gland size by 10%. Further increase in the ADG to 0.8 kg/d had no effect on the mammary gland volume or the milk production. According to Abeni et al. (2000) an average weight gain of 0.9 kg/d significantly reduces the milk fat content. The results presented in Table 5 demonstrate that, in this study, high ADG did not reduce fat or protein content in milk. Heifers with the highest ADG in the rearing period $\geq$ 0.950 kg/d (prepubertal growth $\geq$ 0.970 kg/d) also had the highest components in milk in the first lactation. Hohenboken et al. (1995) argued that the relationship between ADG and milk production depends on the breed and its genetic potential. Smaller breeds are more sensitive to the undesirable effects of intensive rearing in relation to performance in the first lactation (Foldager and Sejrsen, 1991). Kratochvilova (2001) added that growth intensity is highly variable, lowly heritable, and a poor indicator of subsequent milk production. Low correlation between ADG and milk yield found in this study corroborate previous findings.

6.2 Optimal AFC

Views on the optimal AFC are not completely uniform (Mourits et al., 2000). The extension of the rearing period generally caused a negative increase of age (and weight) at the first conception and calving (Stevenson et al., 2008). In this study, as expected, the lowest AFC was achieved with the highest ADG (Table 5). The difference between high ADG $\geq$ 0.950 kg/d and low ADG $\leq$ 0.849 kg/d in the length of the rearing period was 22 d. According to Heinrichs and Gabler (2003), the AFC is influenced by events around the birth as well as nutrition, health, and environmental factors operating during the first 4 mo of life.

Mourits et al. (2000) found that increasing the ADG of 0.9 kg/d in prepubertal heifers and maximum achievement of ADG of 1.1 kg/d during the postpubertal period enables to reach the first calving at 20.5 mo of age with 563 kg body weight, which equalled feed savings of $107/heifer per yr. Our results indicate that approximate AFC of 23.5 mo of age could be achieved by a total ADG higher than 0.950 kg/d (5 mo to 14 mo of age) (Table 5). Mourits et al. (1999b) found no disadvantages of early calving before 22 mo of age, when the growth was maintained and controlled in the heifers' rearing period. Ettema and Santos (2004) added that early calved heifers have lower yields on the first lactation, but higher lifetime performance. Results in Table 5 point to a lower milk yield in the first lactation due to higher ADG (prepubertal growth ≥ 0.970 kg/d and total growth ≥ 0.950 kg/d) compared with medium growth, but animals with higher ADG in the rearing period had a higher milk yield in the second and third lactations than animals with medium or low ADG. Highest LP occurred with groups of prepubertal growth ≥ 0.970 kg/d and total growth ≥ 0.950 kg/d. This finding is consistent with the fact that high ADG does not negatively affect the LP. Similar results were found by Dawson and Carson (2004), who stated that heifers that calved at 540 kg were more economical than those at 620 kg. Late-calved heifers produced 11% more milk in the first lactation, and early-calved heifers lost more weight in early lactation and their calving interval was extended. Nonetheless this fact had no effect on milk production in the second and third lactation. Froidmont (2013) confirmed the strong positive relationship between body weight at the first calving and milk yield in the first lactation. According to Shamay et al. (2005), Holstein heifers should weigh on average 620 kg during the first mo after calving for maximum performance in the first lactation. In this study, the group with the highest BW ≥ 420 kg at the first insemination achieved the highest milk yield in the first, second and third parity (Table 6). According to Hoffman et al. (1996), intensive growth with low efficiency of insemination may lead to negative consequences. Intensive rearing of heifers can lead to calving at 22 mo of age. However, heifers will have undesirable higher BW and BCS at first calving if they are not pregnant until 15 to 16 mo. These heifers will be extremely susceptible to ketosis, displacement of abomasum and low feed intake in the transition period. In this study, a high ADG (≥ 0.950 kg/d) in postpubertal growth negatively influenced the reproduction parameters (Table 5), and high ADG in this period of growth was associated with higher BCS. The BCS from 11 to 16 mo of age (postpubertal period) ranged between 3.2 and 3.5 points. The BCS at 17 mo of age reached above 3.5 points in the group of AFC ≥751 d. One of the reasons for increased BCS in the groups of AFC ≥750 d was because heifers were pregnant (Figure 8). The increased fat reserves in the bodies of the heifers could be one of the reasons for the extended AFC. It might have prolonged the age at first conception and subsequent AFC in the group with AFC ≥751 d. The results also showed that the heifers with low BCS (≤ 3 point) in the conception period (14 mo of age) had reduced reproduction ability in the first parity. However, their calving interval in the second parity was the shortest, 407 d, in comparison to the other groups (Table 5).

Shamay (2005), Ettema and Santos (2004), Heinrichs and Gabler (2003) and Mourits et al. (1999a) suggested that the an AFC in the Holstein breed of 23-24 mo is optimal for maximum profitable production. Ettema and Santos (2004) found in analyses of age and BW at the first calving for Holsteins that only 2.7% of dairy farms achieve the recommended targets and therefore, this leads to economic losses. Wathes et al. (2008) reported that optimal fertility and maintenance of maximum performance in the first lactation were reached at the calving age of 24 to 25 mo, although heifers that calved at the age of 22 to 23 mo were the best in overall performance and longevity over 5 years, partly because heifers with good fertility also had a high level of fertility as cows. Abeni et al. (2000) and Van Amburgh et al. (1998) concluded that calving earlier than 23 mo of age is associated with lower milk yields and lower milk fat content, although on the other hand it also leads to a higher milk protein content, which is influenced by the reduction in milk production. Moreover, they also concluded that earlier calving also leads to reduced reproduction performance. In our study (Table 5), we did not observe a significant impact of AFC on milk yields except on milk yield in the first 100 d of first lactation (P <0.05). We found highest milk yield in the second lactation, third lactation, and LP for the group with AFC ≤699.

6.3 Economic aspects of heifer-rearing period

AFC is an important factor in the cost of rearing replacements in dairy herds and can be adjusted by altering growth rates (Curran et al., 2013; Froidmont et al., 2013). The average AFC in Holstein cattle has been recommended to be ≤24 mo while achieving body size that is adequate to maximize lactation performance, yet control rearing costs (Shamay et al., 2005; Tozer and Heinrichs, 2001; Abeni et al., 2000). The group with lowest AFC (≤749 d) showed the lowest total cost per calf at 6 mo of age of CZK 10,233 (Table 8). Ettema and Santos (2004) have reported that only few dairy farms achieved the recommended AFC target. In fact, the lowest AFC (≤749 d) was achieved by just 8 farms from the evaluated set (Tables 7 and 8). Even when heifers are managed and fed similarly to achieve similar growth rates, variability in AFC is observed. This is dictated by reproduction efficiency at breeding. Poor reproduction increases variability in AFC, even though nutrition and growth rates may be adequate (Ettema and Santos, 2004). Reaching the goal of AFC ≤ 24 mo requires an ADG of between 0.7 and 0.8 kg in Holstein cattle (Abeni et al., 2000). The group with lowest AFC (≤ 749 d ≈ 24.5 mo) had mean ADG of 0.89 kg (Table 7). Shamay et al. (2005) reported that ADG of about 0.7 kg is optimal for achieving maximum performance. The highest MY (8,129 kg) was found in the middle ADG group (0.799 to 0.700 kg) (P < 0.05), and this group of herds also achieved the highest level of profitability, at 2.42% (Table 8). Conversely, the lowest MY (7,327 kg) was achieved by the group having the lowest ADG (≤0.699 kg) (Table 7). That lowest ADG group did, however, had the highest success in conception rate, at about 62%. Hohenboken et al. (1995) had contended that the relationship between ADG and milk production depends on genetic potential. In our study, the regression coefficient and correlation for MY and total ADG were 1,860 kg (P < 0.05) and 0.16 (P < 0.05), respectively. Mourits et al. (2000) had remarked that the optimal AFC is far from uniform and stable. Wathes et al. (2008) found that optimal fertility and maintaining maximum performance was achieved in the AFC range of 24 to 25 mo. The group with lowest AFC (≤ 749 d ≈ 24.5 mo) showed the lowest fertility compared to the other evaluated groups (P < 0.05) (Table 7). Conception rates after first and all inseminations were about 56% for heifers and about 34% for cows. The longest periods were also found in this group for days open (132 d) and calving interval (416 d). It turns out that rearing heifers intensively, which leads to lower AFC, is an important measure for monitoring and evaluating the overall management of a dairy herd inasmuch as it prevents low fertility. The highest culling rate for cows of 41% (P < 0.05), and consequently the highest cow depreciation costs of 8,275 CZK per cow (P < 0.05), and the lowest level of profitability (−9.66%) were found for the same − lowest − group of AFC. Farmers assume that decreasing AFC will reduce costs of rearing heifers and increase their profits because these will produce for longer (Mourits et al., 1999; Van Amburgh et al., 1998). However, the group with the lowest AFC (≤ 749 d ≈ 24.5 mo) in this study had, actually, the lowest number of completed lactations (P < 0.05) and the lowest level of profitability. Honarvar et al. (2010) stated that short herd life leads to high replacement costs and limits breeding selection. Hoffman et al. (1996) considered that lower AFC had to be evaluated in economic terms for each farm on the grounds that low AFC does not always lead to the most profitable solution. Heikkila et al. (2008) found that the variability in results from several studies examining optimal AFC depended upon local conditions and the dairy herd management on each farm. Curran et al. (2013) stated that to make a final economic evaluation of shorter rearing period it is important to know the biological interrelationships between growth rate and subsequent reproduction and between growth rate and the ability for milk production.

6.4 High-production cows and profitability

A few years ago, numerous authors confirmed there to be decreasing fertility in high-producing dairy cows (De Vries et al., 2005; Gonza'lez-Recio et al., 2004; Royal et al., 2000). In this study, the high-producing herds were shown to have the lowest conception rates (Table 7) among cows after first and all inseminations. However, shortest days open (106 d) and calving interval (386 d) were also observed for the same group. Kvapilík et al. (2012) found that higher culling rate lead to lower calving interval, but without positive economic impacts. Culling rate of high-producing herds was the highest. The level of reproductive performance directly affects the economic performance of a dairy herd (Lee et al., 2007). Nonetheless, the group of high-producing herds was the most profitable in this study. Mourits et al. (1999) had concluded that price movements for milk and production inputs bore large income effects because management practices can only partially adjust to these changes. Kvapilík et al. (2012) reported that even as the average farm milk price in the Czech Republic during 2011 was 8.26 CZK the average of total cost of cows was 8.55 CZK per litre of milk and the average of cost of cows less the costs of the by-products (i.e., costs of rearing calves, costs of manure disposal) was 8.39 CZK per litre of milk. It is evident that dairy farms in the Czech Republic would be operating at a loss without subsidies (Table 8). Heikkilla et al. (2008), too, concluded that price movements for milk and production inputs significantly affect farm profitability. Therefore, an optimal replacement policy does not ensure a dairy herd's good economic performance. Nevertheless, Stevenson et al. (2008) agreed with the statement that the level of rearing heifers is one of the most important factors having a significant effect on subsequent reproduction performance and profitability in dairy herds. In this study, the highest-producing herds achieved the lowest AFC, at 780 d ($P < 0.05$), but also the highest culling rate of 38% ($P < 0.05$). This resulted in this group having the highest cow depreciation costs. Lucy et al. (2001) shared the general opinion that increase in milk output is often connected with reduced reproductive ability. However, some studies do not confirm this view and consider that the problem lies in inadequate management and environmental conditions (especially the quality of feedstuffs) in high-producing herds (Chiumia et al., 2013; Kadokawa, 2006). The group of high-producing herds in this study showed the lowest death rate for calves (2.43%) and second-lowest total loss of calves (8.33%). The reason for this probably lies in better nursing care for calves in high-producing herds. Fertility problems were the most common reasons for culling of cows and heifers. As noted by Kadokawa (2006), animals have certain biological limits and any disruption of homeostasis in their bodies leads to difficulties in their future performance.

7 Conclusions and practical implications

The aim of this study was to define connections between heifer rearing intensity (average daily weight gain, age at first calving, body condition scores during the rearing phase) and production performance, fertility and longevity of dairy cows in the production period with regard to overall herd economics.

1) Evaluation of heifer growth and subsequent production period

Control and management of first calving age is a key factor to ensure good lifetime milk performance and ability of cows produce more milk during the production period. The lowest average milk yield per lactation was observed in the group cows with AFC over 24.5 month. The highest ADG in all the groups under study was in heifers with first calving age under 23.5 months. The highest average daily weight gain ($\geq$ 0.850 kg/day) was associated with first calving age 22 days lower than that in the heifers with low ADG (< 0.849 kg/day). The difference in AFC between the highest and lowest ADG heifers was about one oestrous cycle. However, with ADG $\geq$ 0.850 kg/day the risk of culling at first lactation was increased to over 80%. For milk performance traits no negative impact of low ages at first calving (< 23 months) was observed, except lower 100 day milk in the first lactation. Reproduction parameters and milk yield in the second and third lactation were the lowest in the group of cows with first calving age over 24.5 months. An increase in first calving age was associated with higher body condition score in the transition period. The results have shown that first calving ages less than 23 months favour successful heifer rearing and subsequent optimum milk performance and reproduction in well-managed herds than later first calving ages as it was the case in the other groups.

2) The evaluation of heifer rearing intensity and economic impact in the production period

The dairy herd is a complex system consisting of two interconnected parts: the cow herd and the replacement heifer herd. Management decisions regarding herd replacement can have a significant impact on the overall farm profitability. Low first calving ages do not always lead to the highest profit, although it is a common heifer rearing management target. Efficiency of heifer rearing is very much influenced by farm conditions. A more intensive heifer rearing (first calving age $\leq$ 24.5 months) may lead to impaired fertility, increased

58

culling rate and reduced profit. Milk production level had a significant impact on levels of profit achieved by the farms under study. Profitability excluding subsidies was the highest in the group with the highest milk production. Probability of good return on investments increases if herd management is based on good heifer rearing. However, optimum herd turnover does not ensure increased farm profitability, which is influenced by changes in input and output prices at the agricultural market to a considerable extent. In conclusion it should be mentioned that the heifer rearing intensity based on medium first calving ages (799 – 750 days) and average daily weight gain of 0.799 – 0.700 kg was found to be the most profitable. These groups showed the highest values of overall farm profit parameters, excluding subsidies, and 2.42% profitability, excluding subsidies, although the rearing costs were highest in these groups. This study results suggest that there is no recommended universal and optimum first calving age. An optimum first calving age differs among dairy herds and depends on genetic potential and, above all, level of management.

The aforementioned hypothesis stating that on the basis of biological connections between heifer rearing intensity and performance and longevity of production cows it is possible to evaluate economic efficiency of heifer rearing techniques, was confirmed and assessed in the sense of the effect of average daily weight gain during the rearing phase, body condition score and body weight at 14 months of age on production and reproduction parameters in the first three lactations, and also in terms of heifer rearing intensity and overall farm profitability.

The results imply that dairy farmers should thoroughly examine their operation as a whole, and then start to decrease AFC age if pertinent. In herds with poor reproductive parameters, general health status, high culling rates, feed management problems, poor stock care, or inferior housing it is not recommended to perform high intensity heifer rearing. Such herds are very likely to incur profit loss with low first calving ages. On the contrary, good herds, after assessment of breed earliness and benefits of high intensity heifer rearing, can achieve high lifetime performance, good fertility and health in the production period, and reach an optimum herd replacement rate. Cost of one day on feed directly influences losses ensuing from the prolongation of rearing phase over the length which is optimal for that particular herd, or else economic benefits of shortened rearing phase. This study provides the summary of most important periods in the heifer rearing phase including heifer rearing management methods to achieve maximum profit in the subsequent cow production period.

8 Annexes

Tab. 1 Basic parameters of a farm

Farm (address, tel., e-mail):

Farm land (ha)		Permanent grassland (ha)	
Production region		Farming: conventional = 1, organic = 2[1]	
Number of stockpeople		Number of stockpeople	
Average numbers (heads)	Cows total	Cow housing: free stall =1, tie stall = 2	
	Cows milked	Breeding: seasonal = 1, all-year-round = 2	
	Dairy replacement heifers	Heifer housing: free stall =1, tie stall = 2	
Milk yield (litres/cow/year)		Heifer rearing technique: barn = 1, barn + pasture = 2	

1) both the options can be given at the same time (1,2).

Tab. 2 Parameters of heifer rearing[1]

1.	2.	3.	4.	5.
Parameter, item	Birth to .... months of age	..... month to.....year of age	Late pregnancy heifers	Heifers total
Heifer DOF (days on feed)				
ADG (kg/ farm)				
Start of rearing — age (months)				
Start of rearing — BW kg/head				
End of rearing — age (months)				
End of rearing — BW kg/head				
Heifer price[3]				
Milk and milk replacers				
Forages				
Concentrates				
Labour costs				
POL and energy				
Veterinary services				
Breeding services				
Depreciation of tangible fixed assets				
Other items[4]				
Vitamins				
Overheads				
Total costs				

1) when monitoring costs of the whole rearing period (e.g. from birth to 6 months of age), fill the column 5 only;
2) adjust according to the classification you have been using –by giving months of pregnancy;
3) price (market price) of heifers at the start of rearing (arrivals);
4) repairs and maintenance, interest on credits, insurance, fees, and other important items can be listed in the empty lines.

Tab. 3 Annual heifer herd turnover per farm

Parameter	Heads	Kg	Thous. CZK
To 1st January, 2010			
Liveborn calves			
Deaths and necessary culls			
Sold to abattoir			
Sold for further rearing			
Transferred to dairy herd (replacements)			
To 31st December, 2010			

Tab. 4 Parameters of reproduction, mortality and necessary culls

Mean age at first insemination (months/days)		Mean age at first calving (months/days)	
Mean BW at first insemination (kg)		Mean BW at first calving (kg, estimated)	
Ø calving interval (days)		Services per conception - heifers	
Conception rate after 1st insemination (%)		Total services to conceptions (%)	
Total number of inseminations per heifer per year		Calves at 3 mo of age per 100 cows	
Dearth and necessary culls (heads)		Deaths and NP ♀ till 6 mo of age (heads)	

Tab. 5 Direct payments, subsidies and other support per farm (thous. CZK)

SAPS payments		Total top-up payments	
Subsidies according to the Act on Agriculture[1]		Payment per dairy cow	
Less Favoured Areas (LFA)		Organic farming	
Other payments and subsidies		Total direct payments and subsidies	

1) National subsidies (disease funds, eradication programmes, milk testing, etc.)

Tab. 6 Effects of selected factors on heifer rearing success (subjective evaluation)[1]

Factor	Score	Factor	Score	Factor	Score	NB
Breed		Nutrition and feeding		Reproduction management		
Organization of work		Pasture		Natural conditions		
Stockpeople		Input prices		Housing		
Health		Growth rate		[2]		

1)   *Score 1 to 5: 1 = small (negligible) effect, 5 =strong effect. Different factors can have the same scores.*
2)   *Add other factors you regards as important.*

9 References

1. Abeni, F.; Calamari, L.; Stefanini, L.; Pirlo, G. (2000): Effects of daily gain in pre- and pospubertal replacement dairy heifers on BCS, body size, metabolic profile and future milk production. Journal of Dairy Science. 83(7):1468-1478.

2. Bach, A; Ahedo, J. (2008): Record keeping and economics of dairy heifers. Veterinary Clinics of North America-Food Animal Practice. 24 (1):117.

3. Bach, A.; Kertz, A. F. (2010): Raising Dairy Replacements Objectively: The Value of Data-Based On-Farm Decisions. Proceedings of the 19th Annual Tri-State Dairy Nutrition Conference. 77-90.

4. Bar-Peled, U.; Robinzon, B.; Maltz, E.; Tahari, H.; Folman, Y.; Bruckental, I.; Voet, H.; Gacitua, H.; Lehrer, A. R. (1997) : Increased Weight Gain and Effects on Production Parameters of Holstein Heifer Calves That Were allowed to Suckle from Birth to Six Weeks of Age. Journal of Dairy Science. 80:2523-2528

5. Beck, P. A.; Gunter, S. A.; Phillips, J. M.; Kreider D. L. (2005): Development of replacement heifers using programmed feeding. Professional Animal Scientist . 21:365-370.

6. Bellman, R. (1957): Dynamic Programming. Princeton University Press. Princeton. NJ.

7. Bethard, G. L.; James, R. E.; McGilliard, M. L. (1997). Effect of rumen-undegradable protein and energy on growth and feed efficiency of growing Holstein heifers. Journal of Dairy Science. 80:2149-2155.

8. Bicalho, R. C.; Galva˜o, K. N.; Cheong, S. H.; Gilbert, R. O.; Warnick, L. D.; Guard, C. L. (2007): Effect of stillbirth on dam's survival and reproduction performance in Holstein dairy cows. Journal of Dairy Science. 90: 2797–2803.

9. Blome, R.; Drackley, J. K.; McKeith, F. K.; Hutjens, G. C. (2003): Growth, nutrient utilization, and bodycomposition of dairy calves fed milk replacers containing different amounts of protein. Journal of Animal Science. 81:1641-1655.

10. Bouška, J.; et al. (2006): Chov dojeného skotu. Profi Press, s. r. o. 1. vyd. 186 s. ISBN 80-86726-16-9.

11. Britt, J.; Painter, J.; Alvarez, F. (1998): Herd investigation - A comparison of milk production during first lactation.Compendium on continuing education for the practicing veterinarian. 20:645.

12. Cabrera, V. E. (2012): A simple formulation and solution to the replacement problem: A practical tool assess the economic cow value, the value of a new pregnancy, and the cost of a pregnancy loss. Journal od Dairy Sciences. 95:4683-4698.

13. Cabrera, V. E. (2010): A large Markovian linear program to optimizme replacement policies and dairy herd net income for diets and nitrogen excretion. Journal od Dairy Sciences. 93:394 -406.

14. Capuco, A. V.; Smith, J. J.; Waldo, D. R.; Rexroad, C. E.(1995): Influence of Prepubertal Dietary Regimen on Mammary Growth of Holstein Heifers. Journal od Dairy Sciences. 78:2709-2725.

15. Choi, Y. J.; Han, I. K.; Woo, J. H.; Lee, H. J.; Jang, K.; Myung, K. H.; Kim, Y. S. (1997): Compensatory growth in dairy heifers: the effect of a compensatory growth pattern on growth rate and lactation performance. Journal of Dairy Science. 80(3):519-24.

16. Cue, R. I., Pietersma, D., Lefebvre, D., Lacroix, R., Wade, K., Pellerin, D., de Passillé, A-M. and Rushen, J. (2012): Growth modeling of dairy heifers in Québec based on random regression. Canadian Journal Animal Science. 92: 33–47.

17. Daccarett, M. G.; Bortone, E. J.; Isbell, D. E.; Morrill, J. L.; Feyerherm, A. M. (1993): Performance of Holstein heifers fed 100% or more of National Research Council requirements. Journal of Dairy Science. 76:606-614.

18. Daniels, K. M. (2010): Dairy Heifer Mammary Development. Proceedings of the 19th annual tri-state dairy nutrition conference. 69-76.

19. Dawson, L. E. R.; Carson, A. F. (2004): Management of the dairy heifer. Cattle Practice.12:181-192.

20. Demeter, R. M.; Kristensen A. R.; Dijkstra, A. G.; Oude Lansink, A. G. J. M.; Meuwissen, M. P. M.; van Arendonk, J. A. M. (2011): A multi-level hierarchic Markov process with Bayesian updatingfor herd optimization and simulation i dairy cattle. Journal of Dairy Science. 94:5938 - 5962.

21. De Vries, A. (2004): Economics of delayed replacement when cow performance is seasonal. Journal of Dairy Science. 87:2947-2958.

22. De Vries, A.; Risco, C. A. (2005): Trends and seasonality of reproductive performance in Florida and Georgia dairy herds from 1976 to 2002. Journal of Dairy Science. 88 (9):3155 -3165.

23. De Vries, A. (2006): Economic value of pregnancy in dairy cattle. Journal of Dairy Science. 88:3876-3885.

24. Dijkhuizen, A. A. (1992): Modelling animal health economics. Inaugural speech. Wageningen Agricultural University. Wageningen. 28 pp.

25. Dijkhuizen, A. A.; Morris, R. S. (1997): Animal Health Economics – Principles and Applications. Post Graduate Foundation in Veterinary Science – University of Sydney. 1. vyd. 306 s. ISBN 0-646-31481-5.

26. Doležal, O., et al. (2001): Odchov telat ve 222 otázkách a odpovědích. Praha-Agrospoj. 138 s. ISBN 80-239-4228-X.

27. Domecq, J. J.; Skidmore, A. L.; Lloyd, J. W.; Kaneene, J. B. (1995): Validation of body condition scores with ultrasound measurements of subcutaneous fat of dairy cows. Journal of Dairy Science. 78(10):2308-13.

28. Domecq, J. J.; Skidmore, A. L.; Lloyd, J. W.; Kaneene, J. B. (1997): Relationship Between Body Condition Scores and Conception at First Aricial Insemination in a Large Dairy Herd. Journal of Dairy Science. 80:113 – 120.

29. Drevjany, L; Kozel, S; Padrůněk, V. (2004): Holštýnský svět. 1. vyd. Zea Sedmihorky, s. r. o. ve spolupráci se Zemědělským týdeníkem, 344 s.

30. Edmonson, A. J.; Lean, I. J.; Weaver, L. D; et al. (1989): A body condition scoring chart for Holstein dairy-cows. Journal of Dairy Science. 72(1): 68-78.

31. Eicker, S.; Fetrow, J. (2003): New tools for deciding when to replace used dairy cows. Pages 33-46 in Proc. 2003 Kentucky Dairy Conf. Cave City, KY. Univ. Kentucky Lexington.

32. Ettema, J. F.; Santos, J. E. (2004): Impact of age at calving on lactation, reproduction, health, and income in first-parity Holsteins on commercial farms. Journal of Dairy Science. 87:2730–2742.

33. Ford, J. A., Park, C. S. (2001): Nutritionally directed compensatory growth enhances heifer development and lactation potential. Journal of Dairy Science. 84(7):1669-78.

34. Foldager, J.; Krohn, C. C. (1994): Heifer calves reared on very high or normal levels of whole milk from birth to 6-8 weeks of age and their subsequent milk production. Proceedings of the Nutrition Society. 7:1669-1678.

35. Foldager, J., Sejrsen, K. (1991): Rearing intensity in dairy heifers and the effect on subsequent milk production. Institute of Animal Science. No. 693, Denmark.

36. Fricke. P. M. (2003): Heifer Reproduction. In: Raising Dairy Replacements. Midwest Plan Service. Ames, IA, pp. 77-83.

37. Gabler, M. T.; Tozer, P. R.; Heinrichs, A. J. (2000): Development of a Cost Analysis Spreadsheet for Calculating the Costs to Raise a Replacement Dairy Heifer. Journal of Dairy Science. 83:1104-1109.

38. Gasser, C. L.; Behlke, E. J.; Grum, D. E. (2006): Effect of timing of feeding a high-concentrate diet on growth and growth and attainment of puberty in early-weaned heifers. Journal of Animal Science. 84:3118-3122.

39. Groenendaal, H.; Galligan, D.; Mulder, H. (2004): An economic spreadsheet model to determine optimal breeding and replacement decisions for dairy cattle. Journal of Dairy Science. 87:2146–2157.

40. Hansen, M.; Misztal, I.; Lund, M. S.; Pedersen, J.; Christensen, L. G. (2004): Undesired phenotypic and genetic trend for stillbirth in Danish Holsteins. Journal of Dairy Science. 87:1477–1486.

41. Heikkila, A. M.; Nousiainen, J. I.; Jauhiainen, L. (2008): Title Optimal replacement policy and economic value of dairy cows with diverse health status and production capacity. Journal of Dairy Science. 91 (6):2342-2352.

42. Heinrichs, A. J., Gabler, M. T. (2003): Dietary Protein to Metabolizable Energy Ratios on Feed Efficiency and Structural Growth of Prepubertal Holstein Heifers. Journal of Dairy Science. 86(1):268-274.

43. Hoffman, P. C. (2009): Potential of Alternative Dairy Replacement Heifer Nutrition Programs to Reduce Economic Cost and Environmental Impact. Advances in dairy technology. 21:217-225.

44. Hoffman, P. (1997): Optimum body size of Holstein replacement heifers. Journal of Animal Science. 75:836-845.

45. Hoffman, P. C., Brehm, N. M., Price, S. G., Prill-Adams, A. (1996): Effect of accelerated postpubertal growth and early calving on lactation performance of primiparous Holstein heifers. Journal of Dairy Science. 79:2024-2031.

46. Hohenboken, D.; Foldager, J.; Jensen, J.; Madsen, P.; Andersn, B. B. (1995): Breed and nutritional effects and interactons on energy intake, production and efficiency of nutrient utilization in young bulls, heifers and lactating cows. Sect. A – Animal Science. 45:92-98.

47. Honarvar, M.; Javaremi, A. N.; Ashtiani, S. R. M.; Banadaki, M. D. (2010): Effect of length of productive life on genetic trend of milk production and profitability: A simulation study. African Journal of Biotechnology. 9:20.

48. Hulsen, J. (2011): Cow sinals – Jak rozumět řeči krav. Proti Press. 98 s. ISBN: 978-8086726441.

49. Hulsen, J.; Swormink, B.K. (2006): From Calf to Heifer: A Practical Guide for Rearing Young Stock. Roodbont Publishers. 40 s. ISBN: 978-9075280951.

50. Hultgren, J.; Svensson, C. (2010): Calving Interval in Dairy Cows in Relation to Heifer Rearing Conditions in Southwest Sweden. Reproduction in Domestic Animals. 45:136-141.

51. Hultgren, J.; Svensson, C. (2009a): Lifetime risk and cost clinical mastitis in dairy cows in relation to heifer rearing conditions in southwest Sweden. Journal of Dairy Science. 92: 3274-3280.

52. Hultgren, J.; Svensson, C. (2009b): Heifer rearing conditions affect length of productive life in Swedish dairy cows. Preventive Veterinary Medicine. 89:255-264.

53. Jasper, J., Weary, D. M. (2002): Effects of Ad Libitum Milk Intake on Dairy Calves. Journal of Dairy Science. 85(11):3054-8.

54. Kadokawa, H.; Martin G. B. (2006): A New Perspective on Management of Reproduction in Dairy Cows: the Need for Detailed Metabolic Information, an

Improved Selection Index and Extended Lactation. Journal of Reproductive Development. 52:161-168.

55. Kalantari, A. S.; Cabrera V. E. (2012): The effect of reproductive performance on the dairy cattle herd value assessed by integrating a daily dynamic programming model with a daily Markov chain model. Journal of Dairy Science. 95:6160-6170.

56. Kalantari, A. S.; Mehrabani-Yeganeh, H.; Moradi, M.; Sanders, H.; De Vries, A. (2010): Determining the optimum replacement policy for Holsteindairy herds in Iran. Journal of Dairy Science. 93:2262-2270.

57. Kertz, A. F.; Reutzel, L. F.; Mahoney, J. H. (1984): Ad Libitum Water Intake by Neonatal Calves and Its Relationship to Calf Starter Intake, Weight Gain, Feces Score, and Season. Journal of Dairy Science. 67(12):2964-9.

58. Knight, C. H.; Sorensen, A. (2001): Windows in early mammary development: critical or not? . Reproduction. 122(3):337-345

59. Knott, L.; Tartlon, J. F.; Craft, H.; Webster A. J. F. (2007): Effects of housing, parturition and diet change on the biochemistry and biomechanics of the support structures of the hoof of dairy heifers. Veterinary Journal. 174 (2): 277-287.

60. Kratochvilova, M. (2001): Relationship between growth and milk production in dairy cattle. Czech J. Anim. Sci. 46 (3):139-144.

61. Kristensen, A. R. (1992): Optimal replacement in the dairy herd. Agricultural Systems. 1992. 39:1-24.

62. Kvapilík, J.; Růžička Z.; Bucek P. (2012): Ročenka. Chovu skotu v České republice - Hlavní výsledky a ukazatele pro rok 2011. ČMSCH a.s. Praha. 52 s. ISBN: 978-80-87633-02-1

63. Kvapilík, J.; Syrůček, J. (2012): Kalkulace příspěvku na úhradu a úplných nákladů. Náš chov, 2012, roč. 72, č. 3, s. 22-26.

64. Kvapilík, J. (2010): Hodnocení ekonomických ukazatelů výroby mléka. Certifikovaná metodika. VÚŽV, v.v.i. Praha. 78 s. ISBN 978-80-7403-059-8.

65. Lammers, B. P.; Heinrichs, A. J. (2000): The Response of Altering the ratio of Dietary Protein to Energy on Growth, Feed Efficiency, and Mammary Development in Rapidly Growing Prepubertal Heifers. Journal of Dairy Science. 83:977-983.

66. Le Cozler, Y.; Lollivier, V.; Lacasse, P.; Disenhaus, C. (2008): Rearing strategy and optimizing first-calving targets in dairy heifers: a review. Animal. 9:1393-1404.

67. Le Cozler, Y.; Peccatte, J. R.; Porhiel, J. Y.; Brunschwig, P.; Disenhaus, C. (2009a): Rearing dairy heifers. Productions Animale. 22:303-316.

68. Le Cozler, Y.; Peyraud, J. L.; Troccon, J. L. (2009b): Effect of feeding regime, growth intensity and age at first insemination on performances and longevity of Holstein heifers born during autumn. Livestock Science. 124:72-81.

69. Lee, J. I.; Kim, I. H. (2007): Pregnancy loss in dairy cows: the contributing factors, the effects on reproductive performance and the economic impact. Journal of Veterinary Science. 8 (3):283-288.

70. Leroy, J. L.; De Kruif, A. (2006): Reduced reproductive performance in high producing dairy cows: is there actually a problem?.Vlaams Diergeneeskundig Tijdschrift. 75 (2A):55 -60.

71. Lucy, M. C. (2001): Reproductive Loss in High-Producing Dairy Cattle: Where Will It End?. Journal of Dairy Science. 84:1277-1293.

72. Macdonald, K. A.; Penno, J. W.; Bryant, A. M.; Roche, J. R. (2005): Effect of feeing level pre- and post-puberty and body weight at first calving on growth, milk production, and fertility in grazing dairy cows. Journal of Dairy Science Journal of Dairy Science. 88(9): 3363-3375.

73. Madgwick, S.; Evans, A. C.; Beard, A. P. (2005): Treating heifers with GnRH from 4 to 8 weeks of age advanced growth and the age at puberty. Theriogenology. 63:2323-2333.

74. Meyer, M. J.; Everett, R. W.; Van Amburgh, M. E. (2004). Reduced age at first calving: effects on lifetime production, longevity, and profitability. Proceedings. 3rd Annual Arizona Dairy Producers Conference. Tempe, AZ. pp. 41-55.

75. Mourits, M. C. M.; Galligan, D. T.; Dijkhuizen, A. A.; Huirne, R. B. M. (2000): Optimization of dairy heifer management decisions based on production conditions of Pennsylvania. Journal of Dairy Science. 83:1989-1997.

76. Mourits, M. C. M.; Huirne, R. B. M.; Dijkhuizen, A. A.; Kristensen, A. R.; Galligan, D. T. (1999a): Economic optimization of dairy heifer management decisions. Agricultural Systems. 61:17-31.

77. Mourits, M. C. M.; Huirne, R. B. M.; Dijkhuizen, A. A.; Galligan, D. T. (1999b): Optimal heifer management decisions and the influence of price and production variables. Livestock Production Science. 60: 45-58.

78. Nor, N. M.; Steeneveld, W.; Van Werven, T.; Mourits, N. C. M.; Hogeveen, H. (2013): First-calving age and first-lactation milk production on Dutch dairy farms. Journal of Dairy Science. 96 (2):981-992.

79. Nordlund, K. V.; Garret, E. F.; Oetzel, G. R. (1999): Výživa a krmení – Acidóza bachoru. Černostrakaté novinky. str. 69-73.

80. Owens, F. N., Dubeski, P., Hanson, C. F. (1993): Factors that afect the growth and development of ruminants. Journal of Animal Science. 71:3138-3150.

81. Park, C. S.; Danielson, R. B.; Kreft, B. S.; Kim, S. H.; Moon, Y. S.; Keller, W. L. (1998): Nutritionally directed compensatory growth and effects on lactation potential of developing heifers. Journal of Dairy Science. 81(1):243-9.

82. Patterson, D. J.; Perry R. C.; Kiracofe G. H.; Bellows R. A.; Staigmiller R. B.; Corah, L. R. (1992): Management considerations in heifer development and puberty. Journal of Animal Science. 70:4018–4035.

83. Perotto, D.; Cue, R. I.; Lee, A. J. (1992): Comparison of nonlinear functions for describing the growth curve of three genotypes of dairy cattle. Canadian Journal of Animal Science. 72(4): 773-782.

84. Philips, C. J. C. (2010): Principles of cattle production. Cambridge University Press. 2nd edit. 233 s. ISBN 978-80-7403-087-1.

85. Pirlo, G.; Miglior, F.; Speroni, M. (2000): Effect of Age at first calving on production trans and on diference between milk yield returns and reraring costs in Italian Holsteins. Journal of Dairy Science. 83:603–608.

86. Poláčková, J.; Boudný, J.; Janotová, B.; Novák, J. (2010): Metodika kalkulací nákladů a výnosů v zemědělství. Certifikovaná metodika. Ústav zemědělské ekonomiky a informací. Praha. 73 s. ISBN: 978-80-86671-75-8.

87. Pryce, J.E.; Coffey, M.P.; Simm, G. (2001): The Relationship Between Body Condition Score and Reproductive Performance. Journal of Dairy Science. 84:1508–1515.

88. Přibyl, J. (1997): Šlechtění skotu a jeho vliv na jednotlivé chovy. Praha. 1.vyd.. Institut výchovy a vzdělávání MZe ČR. 36 s. ISBN 80-7105-155-1.

89. Roberts, A. J.; Geary, T. W.; Grings, E. E.; Waterman, R. C.; MacNeil, M. D. (2009): Reproductive performance of heifers offered ad libitum or restricted access to feed for a one hundred forty-day period after weaning. Journal of Dairy Science. 87:3043-3052.

90. Roche, J. R.; Friggens, N. C.; Kay, J. K.; Fisher, M.W.; Stafford, K.J.; Berry, D. P. (2009): Invited review: Body condition score and its associaton with dairy cow productivity, health, and welfare. Journal of Dairy Science. 92(12): 5769-5801.

91. Roche, J. F.; Mackey, D., Diskin, M. D. (2000): Reproductive management of postpartum cows. Animal Reproduction Science. 60:703 – 712.

92. Sakaguchi, M.; Suzuki, T.; Sasamoto, Y.; Takahashi, Y.; Nishiura, A.; Aoki, M. (2005): Effects of first breeding age on the production and reproduction of Holstein heifers up to the third lactation. Animal Science. 419-426.

93. SAS Institute Inc. (2008) SAS/STAT® 9.2. User's Guide. Cary, NC: SAS Institute Inc.

94. Schröder, U. J.; Staufenbiel, R. (2006): Invited review: Methods to determine body fat reserves in the dairy cow with special regard to ultrasonographic measurement of backfatthickness. Journal of Dairy Science. 89(1):1-14.

95. Sejrsen, K.; Purup, S. (1997): Influence of prepubertal feeding level on milk yield potential of dairy heifers. Journal of Animal Science. 75:828-835.

96. Sejrsen, K.; Purup, S.; Vestergraad, M.; Foldager, J. (2000): High body weight gain and reduced bovine mammary growth: physiological basis and implications for milk yield potential. Domestic Animal Endocrinology. 19:93-104.

97. Shamay, A.; Werner, D.; Moallem, U.; Barash, H.; Bruckental, I. (2005): Effect of Nursing Management and Skeletal Size at Weaning on Puberty, Skeletal Growth Rate, and Milk Production During First Lactation of Dairy Heifers. Journal of Dairy Science. 44 (4): 1460-1469.

98. Silva, L. F. P.; Vandehaar, M. J.; Whitlock, B. K.; Radcliff, R. P.; Tucker, H. A. (2002): Short Communication : Raltionship Between Body Growth and Mammary Development in Dairy Heifers. Journal of Dairy Science. 85:2600-2602.

99. Spiekers, H.; Potthast, V. (2004): Erfolgreiche Milchvieh-fütterung (4. völlig neu überarbeitete Auflage. DLG Verlag. Frankfurt am Main. ISBN 3-7690-0573-2. Seite 285 – 289.

100. Spiekers, H.; Nussbaum, O.; Potthast, V. (2009): Erfolgreiche Milchviehfütterung; 5. erweiterte und aktualisierte Auflage. DLG Verlag. ISBN 978-3-7690-0730-5.

101. Stevenson, J. L.; Rodrigues, J. A.; Braga, F. A.; Bitente, S.; Dalton, J. C.; Santos, J. E. P.; Chebel, R. C. (2008): Effect of breeding protocols and reproductive tract score on reproductive performance of dairy heifers and economic outcome of breeding programs. Journal of Dairy Science. 91:3424-3438.

102. St-Pierre, N. R. (2002): Application of mixed model methodology to the determination of the economic optimal pre-pubertal rate of gain in dairy heifers. Journal of Dairy Science. 85:(Suppl. 1), 42. (Abstr.).

103. Svensson, C.; Nyman, A.- K.; Persson Waller, K.; Emanuelson, U. (2006): Effects of Housing, Management, and Health of Dairy Heifers on First-Lactation Udder Health in Southwest Sweden. Journal of Dairy Science. 89 (6):1990-1999.

104. Synek, M.; et al. (2011): Manažerská ekonomika. Grada Publishing, a. s. 5. vyd. 471 s. ISBN 978-80-247-3494-1.

105. Tozer, P. R.; Heinrichs, A. J. (2001): What Affects the Costs of Raising Replacement Dairy Heifers: A Multiple-Component Analysis. Journal of Dairy Science. 84:1836-1844.

106. Uys, J. L.; Loučena, D. C.; Thompson, P. N. (2011): The effect of unrestricted milk feeding on the growth and health of Jersey calves. Journal of the South African Veterinary Association. 82(1):47-52.

107. Vacek, M.; Krpálková, L.; Zink, V.; Janecká, M. (2013):Metodika řízení odchovu a reprodukce jalovic holštýnského plemene z hlediska celkové rentability chovu dojnic. VÚŽV, v.v.i. Praha. 31 s. ISBN 978-80-7403-107-6.

108. Van Amburgh, M. E.; Galton, D. M.; Bauman, D. E.; Everett, R. W.; Fox, D. G. L; Chase, E.; Erb, H. N. (1998): Effects of three prepubertal body growth rates on performance of Holstein heifers during first lactation. Journal of Dairy Science. 81:527-538..

109. Veauthier, G., et al. (2000): Intensive Färsenaufzucht (Top agrar Fachbuch); Neuauflage. München. ISBN: 3-7843-3046-0.

110. Verbeke G.; Molenberghs, G. (2000): Linear Mixed Models for Longitudinal Data. Springer-Verlag. New York. 568 s.

111. Waldo, D. R.; Capuco, A. V.; Rexroad, C. E. (1998): Milk Production of Holstein Heifers Fed Either Alfalfa or Corn Silage Diets at Two Rates of Daily Gain. Journal Of Dairy Science. 81:756-764.

112. Wathes, D. C.; Brickell, J. S.; Bourne, N. E.; Swali, A.; Cheng, Z. (2008): Factors influencing heifer survival and fertility on commercial dairy farms. Animal. 8:1135-1143.

113. Wattiaux, M. A. (2011): Heifer Raising - Birth to Weaning. Chapter 35: Measuring Growth. The Babcock Institute for International Dairy Resaerch and Development [online]. © 1994-2011 Board of Regents of the University of Wisconsin System. Dostupné z <http://babcock.wisc.edu/node>.

114. Widiati, R.; Adiarto, A.; Hertanto, B. S. (2012): Profitability of smallholder dairy farms based on the performance of lactating cows and fresh milk market prices at lowland areas of Yogyakarta. Journal of the Indonesian Tropical Animal Agriculture. 37(2): 132-138.

115. Zanton, G. I.; Heinrichs, A. J. (2005): Meta-analysis to assess effect of prepubertal average daily gain of Holstein heifers on first-lactation production. Journal of Dairy Science. 88:3860-3867.

116. Zeman, L., et al. (2006): Výživa a krmení hospodářských zvířat.Profi Press.1 vyd.. 360 s.. ISBN 80-86726-17-7.

in form
on the farm

WWW.VVS.CZ

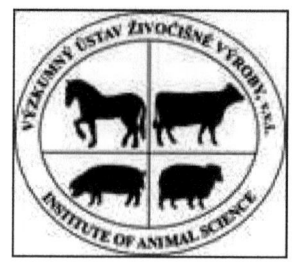

WWW.VUZV.CZ

More Books!

i want morebooks!

Buy your books fast and straightforward online - at one of the world's fastest growing online book stores! Environmentally sound due to Print-on-Demand technologies.

Buy your books online at

www.get-morebooks.com

Kaufen Sie Ihre Bücher schnell und unkompliziert online – auf einer der am schnellsten wachsenden Buchhandelsplattformen weltweit!
Dank Print-On-Demand umwelt- und ressourcenschonend produziert.

Bücher schneller online kaufen

www.morebooks.de

OmniScriptum Marketing DEU GmbH
Heinrich-Böcking-Str. 6-8
D - 66121 Saarbrücken
Telefax: +49 681 93 81 567-9

info@omniscriptum.de
www.omniscriptum.de

Printed by Books on Demand GmbH, Norderstedt / Germany